Lecture Notes in Bioengineering

Stefano Piotto · Federico Rossi
Simona Concilio · Ernesto Reverchon
Giuseppe Cattaneo
Editors

Advances in Bionanomaterials

Selected Papers from the 2nd Workshop in Bionanomaterials, BIONAM 2016, October 4–7, 2016, Salerno, Italy

 Springer

Editors
Stefano Piotto (iD)
Department of Pharmacy
University of Salerno
Fisciano
Italy

Federico Rossi
Department of Chemistry and Biology
University of Salerno
Fisciano
Italy

Simona Concilio (iD)
Department of Industrial Engineering
University of Salerno
Fisciano
Italy

Ernesto Reverchon (iD)
Department of Industrial Engineering
University of Salerno
Fisciano
Italy

Giuseppe Cattaneo
Department of Informatics
University of Salerno
Fisciano
Italy

ISSN 2195-271X
Lecture Notes in Bioengineering
ISBN 978-3-319-87216-2
DOI 10.1007/978-3-319-62027-5

ISSN 2195-2728   (electronic)

ISBN 978-3-319-62027-5   (eBook)

Printed on acid-free paper

This Springer imprint is published by Springer Nature
The registered company is Springer International Publishing AG
The registered company address is: Gewerbestrasse 11, 6330 Cham, Switzerland

Preface

This volume of the Springer book series Lecture Notes in Bioengineering includes the proceedings of the 2nd Workshop on Bionanomaterials, BIONAM 2016, held on October 5–6, 2016, in Salerno, Italy. BIONAM has been focusing on the analysis, synthesis and design of bionanomaterials, with its first edition held in 2013 in Salerno. After the success of this edition, the second edition has been jointly organized with WIVACE 2016, a Workshop on Artificial Life and Evolutionary Computation, with the purpose of involving (more) research on multidisciplinary and experimental fields such as systems chemistry and biology, and emerging topics such as origin of life as well as chemical and biological smart networks. BIONAM merges the biophysicists', the biochemists' and bioengineers' perspectives, covering the study of the basic properties of materials and their interaction with biological and environmental systems, the development of new devices for medical purposes such as implantable systems, and new algorithms and methods for modeling the mechanical, physical or biological properties of biomaterials. This challenging task requires powerful theoretical and computational tools to understand and control the inherent complexity of the interactions between synthetic and biological objects.

The interaction between the WIVACE and the BIONAM communities resulted in a joint session where the experimental work has been harmonized in a well-established theoretical framework; some selected contributions, having a more theoretical character, have been collected in the part *Modelling of Bionanomaterials* of this volume. The volume includes two further parts: *Nanomaterials Engineering* and *Applications of Bionanomaterials*, both accounting for more applicative and technical aspects of nanomaterials.

The BIONAM 2016 volume reports on selected experimental contributions and computational ones, derived from a close interaction between chemical scientists and the bioengineers involved in bionanomaterials studies.

Events like BIONAM are a good opportunity for both new generation and young scientists: to get in touch with new subjects and bring new ideas to the attention of senior researchers; and to highlight and promote the work of the youngest participants. At this purpose, we awarded Dr. Emiliano Altamura and Dr. Francesco Milano for the best poster presentation. Their contribution was selected as full papers and appear in this volume under the title *Modelling Giant Lipid Vesicles designed for Light Energy Transduction*.

As Editors, we wish to express gratitude to all the conference attendees and chapters' authors. We also acknowledge the priceless work of the reviewers and of members of the Program Committee. Special thanks go to the invited speakers **Nicola Tirelli** from the University of Manchester (England) and **Gaetano Guerra** from the University of Salerno (Italy), for their very interesting and inspiring talks.

The 15 papers included in this book have been thoroughly reviewed and selected out of 34 submissions. By covering topics such as *evolutionary computation, bioinspired algorithms, genetic algorithms, bioinformatics and computational biology, modelling and simulation of artificial and biological systems, complex systems, synthetic and systems biology* and *systems chemistry*, they represent the most interesting contributions to the 2016 edition of BIONAM.

Salerno, Italy
October 2016

Stefano Piotto
Federico Rossi
Simona Concilio
Ernesto Reverchon
Giuseppe Cattaneo

Organization

BIONAM 2016 has been organized in Fisciano (SA, Italy) by the University of Salerno (Italy).

Chairs

Stefano Piotto   Department of Pharmacy, University of Salerno (Italy). piotto@unisa.it

Simona Concilio   Department of Industrial Engineering, University of Salerno (Italy). sconcilio@unisa.it

Federico Rossi   Department of Chemistry and Biology, University of Salerno (Italy). frossi@unisa.it

Program Committee

Barba Anna Angela	University of Salerno
Bevilacqua Vitoantonio	Politecnico di Bari
Caivano Danilo	University of Bari
Cattaneo Giuseppe	University of Salerno
Concilio Simona	University of Salerno
Damiani Chiara	University of Milan Bicocca
Mavelli Fabio	University of Bari
Palazzo Gerardo	University of Bari
Pantani Roberto	University of Salerno
Piotto Stefano	University of Salerno
Reverchon Ernesto	University of Salerno
Roli Andrea	University di Bologna

Rossi Federico	University of Salerno
Serra Roberto	University of Modena and Reggio
Stano Pasquale	University of Roma Tre
Villani Marco	University of Modena and Reggio

Supported By

Contents

Part I
Nanomaterials Engineering

Nanoliposomes Production by a Protocol Based on a Simil-Microfluidic Approach

Sabrina Bochicchio, Annalisa Dalmoro, Federica Recupido,
Gaetano Lamberti and Anna Angela Barba

Abstract In this work a protocol based on the microfluidic principles has been developed and applied to produce nanoliposomes. The protocol basically consists in the realization of a contact between two flows, lipids/ethanol and water solutions, inside a tubular device where interdiffusion phenomena allow the formation of lipid vesicles. Effects of solutions flow rates and lipids concentrations on size and size distribution have been investigated. Moreover, ultrasonic energy was used to enhance homogenization of the hydroalcoholic final solutions and to promote the vesicles size reduction. By this protocol a massive output has been achieved; increasing the ratio between the water volumetric flow rate to the lipids-ethanol volumetric flow rate the liposomes dimension decreases; at equal flow rates, when the lipids concentration increases also the liposomes size has been observed increasing.

keywords Liposomes · Ultrasonic energy · Semicontinuous bench scale apparatus

1 Introduction

Lipid-based drug delivery systems are biocompatible carriers even more investigated by the scientific world for their ability in encapsulating and releasing, in a controlled manner, degradable active ingredients to be used for pharmaceutical and nutraceutical purposes. In particular the demand of liposomes based products to be used in several branches of biotechnology has attracted a lot of attention for the liposomes biodegradability, high drug loading capacity, low intrinsic toxicity, accumulation in pathological areas, reduced size, membrane mimetic behavior,

S. Bochicchio · F. Recupido · G. Lamberti
Department of Industrial Engineering, University of Salerno,
Via Giovanni Paolo II 132, 84084 Fisciano, SA, Italy

S. Bochicchio · A. Dalmoro · F. Recupido · A.A. Barba (✉)
Department of Pharmacy, University of Salerno, Via Giovanni Paolo II 132,
84084 Fisciano, SA, Italy
e-mail: aabarba@unisa.it

© Springer International Publishing AG 2018
S. Piotto et al. (eds.), *Advances in Bionanomaterials*,
Lecture Notes in Bioengineering, DOI 10.1007/978-3-319-62027-5_1

3

prolonged half-life in the bloodstream, low cost and easiness of preparation [1]. Size and size distribution are key parameters determining liposomes performance as carrier systems in both biomedical applications (i.e., influencing liposomes time of circulation in the blood stream and/or their permeability through membrane fenestration in tumour blood vessels [2]), cosmetic and nutraceutical applications (i.e., improving consistence, stability, taste, flavor, absorption and bioavailability of nutraceuticals [3, 4]).

Nowadays there is a wide set of possibilities to produce lipid-based drug delivery systems through the use of conventional or more recently discovered techniques [5–7]. However, despite the leaps and bounds made with the novel technologies in the last few years, the majority of these methods are characterized by high energy request, long times of process together with a low productivity.

To overcome these limitations, in this study microfluidics based methods, which are expensive for special devices needed and microfabrication costs, have been transposed to a millimeter scale, drastically reducing the production costs and increasing the yields. The developed simil-microfluidic method provides a new platform in the optimized productions of liposomes. Compared to traditional bulk methods, as Thin Film Hydration (TFH) and Ethanol Injection (EI) techniques, in which heterogeneous and uncontrolled lipids self-assembly processes take place, the simil-microfluidic approach gives the possibility to control liposomes size and size distribution by varying several parameters (i.e., volumetric flow rates ratio, lipids concentration) which controlling the lipids solvent/water interdiffusion phenomena. Moreover, ever respect to the conventional techniques as the TFH, which allow the formation of micrometric structures, the simil-microfluidic method gives liposomal structures directly on nanometric scale. Finally, if compared to the slow and discontinuous processes characterizing traditional methods, the realized simil-microfluidic set up has allowed a semi-continuous and rapid liposomes production. In particular in this work the simil-microfluidic bench scale apparatus has been used for a massive nanoliposomes production. This has been achieved by coupling the ethanol injection method with the microfluidic principles and transposing them on a micrometric scale in an easy and cheap way. In that regard, in a work of Pradhan and collaborators [8] a syringe pump driven microfluidic device was developed to produce nanoliposomes. In the present work the semicontinuous bench scale apparatus was designed by replacing the syringe pumps by the peristaltic once, allowing the production of larger liposomes volumes and by injecting the lipid phase directly into the polar phase without the help of any tubes connection systems as done in [8]. Ultrasonic energy was used as intensification tool for liposomes suspensions homogenization [9]. Furthermore, similarly as done by Jahn and collaborators for a microfluidic hydrodynamic focusing (MHF) platform [10], a size and size distribution control of the produced nanoliposomes was demonstrated by tuning not only the flow rates, as done by Jahn research group, but also the lipids concentration.

2 Experimental

2.1 Materials and Methods

Liposomes were formulated with L-α-Phosphatidylcholine (PC) from soybean, Type II-S, 14–23% choline basis (CAS n. 8002-43-5), purchased from Sigma Aldrich (Milan, Italy) and ethanol of analytical grade (CAS n. 64-17-5, Sigma Aldrich) was used to solubilize the PC. Rhodamine B was used for microscope fluorescence observations of the produced liposomes.

In Fig. 1 a schematic representation of the used experimental apparatus is presented. The first part of the bench-scale apparatus consists in two containers, one filled with a lipids/ethanol solution (prepared by dissolving 16.5 mg of PC in 10 ml of ethanol), which is conveyed in a 1.6 mm silicon tube, and one filled with the hydration solution (distilled water) conveyed in a 5 mm silicon tube through the use of two peristaltic pumps (Verderflex OEM mod. Au EZ). The lipids/ethanol solution tube ends with a needle (0.6 mm internal diameter) inserted into the production section tube, a 3 mm internal diameter silicon tube which is an extension of the water tube. This is the production section, where a diffusive mechanism of the two pushed liquids takes part leading to the formation of liposomes during the two phases (water and alcohol) interdiffusion. The hydroalcoholic solution containing liposomes on nanometric scale is recovered in a container and then subjected to a duty cycle sonication process in order to homogenize and to further reduce vesicles size.

At first, liposomes were prepared by maintaining constant the PC concentration in the hydroalcoholic solution at 0.15 mg/ml and varying the volumetric flow rates

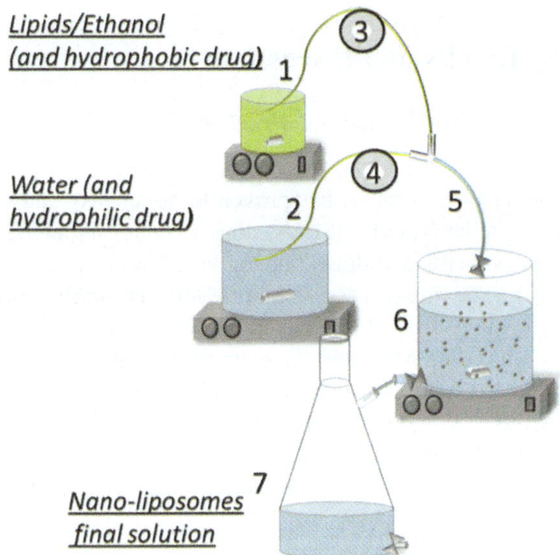

Fig. 1 Schematization of the simil-microfluidic apparatus. From the containers (1–2) lipids/ethanol and water solutions are pushed through peristaltic pumps (3–4) to the production section (5) were vesicles are formed. The hydroalcoholic solution is recovered in a container (6) where it is stirred. The nanoliposomes solution is finally subjected to duty cycle sonication and recovered in a receiving flask (7)

in order to study their influence on liposomes size and size distribution. The volumetric flow rates ratio, defined as hydration solution volumetric flow rate (Vhs) to lipid solution volumetric flow rate (Vls), was varied (10:1, 15:1, 20:1 and 40:1 Vhs/Vls) maintaining constant the Vls at 4 ml/min and changing the Vhs (40, 60, 80 and 160 ml/min). Subsequently, the PC concentration in the final hydroalcoholic solution was changed (0.15, 1, 4 and 5 mg/ml) by maintaining constant at 10:1 (Vhs/Vls) the volumetric flow rate ratio, and the liposomes size and size distribution were analyzed again. Ultrasonic energy was used as intensification tool in homogenization operation through irradiating procedures previously developed and described in [9] by applying a duty cycle consisting of 5 ten-second irradiation rounds each followed by a 20 s pause in order to prevent thermal vesicle disruption (VCX 130 PB Ultrasonic Processors, Sonics & Materials Inc., Newtown, CT, USA). For both the experimental campaigns the volumetric flow rate ratio and PC concentration effects on liposomes size and size distribution were studied with and without the ultrasound contribution.

Morphological observations of liposomes were performed using optical microscopy (Axioplan 2- Image Zeiss) for fluorescence imaging, a 100 X oil immersion objective was used to visualize the vesicles. Dimensional characterization of vesicles was performed by Dynamic Light Scattering analysis (DLS, Zetasizer Nano ZS, Malvern, UK). The resulting particle size distribution was plotted as number of liposomes versus size. The average size refers to the peak size (or distribution size) with a statistical mean and standard deviation of that specific peak. The Polydispersity Index (PDI) was calculated for all the preparations.

Statistical analysis of experimental data (size and PDI) were analyzed using Student's t-test and one-way analysis of variance (ANOVA), and differences were considered statistically significant at $p < 0.05$.

3 Results and Discussion

3.1 Nanoliposomes Production

The developed set up has proven to be an easy and fast method to produce Small Unilamellar Vesicles (SUVs) directly on a nanometric scale, without the use of any toxic solvents and drastic operative conditions (i.e., high temperature and/or pressure) achieving a massive output with the minimum of energy and costs. Results concerning size of produced liposomes have shown vesicles formation of 33–49 nm in size when the volumetric flow rates ratio was varied; of 49–81 nm in size when the lipid concentration was changed (dimensional characterization by DLS analysis). For this reason, the vesicles presented in the fluorescence microscopy images of Fig. 2 are called SUVs even if aggregates and/or large vesicles produced together with SUVs, although few, are well and more visible in the captured images.

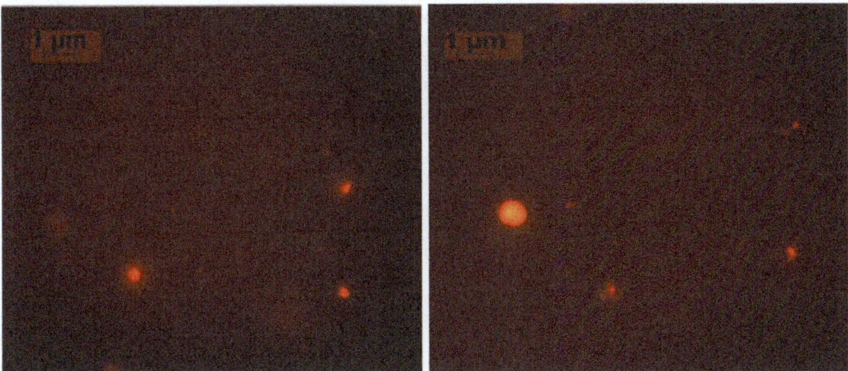

Fig. 2 Fluorescence microscopy images of lipid vesicles labelled with Rhodamine B dye and visualized with a 100 X objective

From a phenomenological point of view, liposomes formation happens at the interfaces between the alcoholic and water phases, when they start to interdiffuse in a direction normal to the liquid flow stream. When lipids, dissolved in the alcoholic phase, meet with water, due to their insolubility, they start to assemble together generating pieces of bilayer which close themselves engulfing water and forming spherical vesicles. Even if the liposomes formation is a spontaneous process, their dimension and size distribution as well as the productivity of the process (the number of liposomes for unit volume obtainable) can be controlled operating on the volumetric flow rates ratio of the two liquids, on lipids concentration and using ultrasonic energy as tool for the process intensification.

3.2 Influence of Volumetric Flow Rate Ratio on Liposomes Formation

As reported in Fig. 3, increasing the ratio between the water volumetric flow rate to the lipids-ethanol volumetric flow rate the liposomes dimension decreases (Fig. 3a) while the PDI value increases (Fig. 3b). In particular, starting with a 10:1 Vhs/Vls ratio, liposomes of 49 nm in size were obtained up to a diameter of 33 nm for liposomes achieved with 40:1 Vhs/Vls, the higher volumetric flow rate ratio tested. Viceversa opposite trend is visible for the PDI whose 0.36 value obtained from the 10:1 Vhs/Vls becomes 0.73 when a 40:1 Vhs/Vls ratio was used, index, the latter, of a highly polydispersed sample. Taking into account that the Vhs/Vls ratio was increased by enhancing the water flow rate and maintaining constant that of the lipids-ethanol solution, what plays a key role in the liposomes diameter size reduction is the diffusion rate of the lipids into the hydration solution. Considering constant and uniform the lipids concentration for unit of volume in the alcoholic

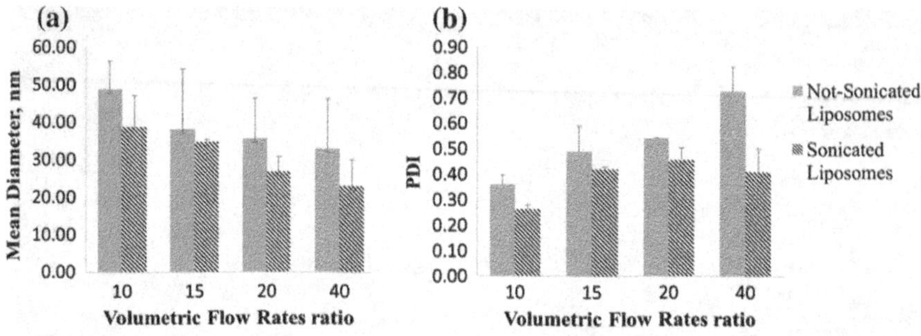

Fig. 3 **a** Sonicated and not sonicated liposomes mean diameter size at different volumetric flow rates ratio. **b** Polydispersity Index (PDI) of sonicated and not sonicated liposomes at different volumetric flow rates ratio. Results are expressed as average of three determinations and reported along with the standard deviation

solution, when the water volumetric flow rate increases the impact between the two liquids increases too and lipids spread faster in a greater water volume. Lipids so scattered join together to form vesicles even more small with the increase of water flow rate. For both size and size distribution it can be observed that the applied ultrasound assisted process did not globally affect, for all the volumetric flow rates explored in this study, the trend found for the not-sonicated samples.

In particular, at constant flow rate ratio there isn't evidence of size changes between sonicated and not sonicated samples (p > 0.05), whilst sonication has effect on PDI. In particular, for not sonicated samples, PDI is worsened when flow rate ratio increases from 20 to 40 (it is kept statistically constant between 10 and 20); for sonicated samples, PDI is worsened going from a flow rate of 10–15; for higher flow rates it can be considered constant (Fig. 3b). Reasonably it is possible to affirm that homogenization assisted by ultrasonic energy can reduce liposomes aggregates.

3.3 Influence of Lipid Concentration on Liposomes Formation

It was observed that at equal flow rates, when the lipids concentration increases also the liposomes size has risen. In Fig. 4a data shown illustrate liposomes diameter average size obtained at different PC concentrations in the hydroalcoholic solution. As the PC concentration increases from 0.15 to 5 mg/ml the liposomes size also increases from approximately 49 to 81 nm. It can be explained by the fact that at higher PC concentrations more phospholipids will impact at the same alcohol/water interface area and will dissolved in the same water volume making them physically closer to each other, thus forming larger vesicles. Homogenization step, performed by sonication, seems do not have effect on liposomes size (p > 0.05).

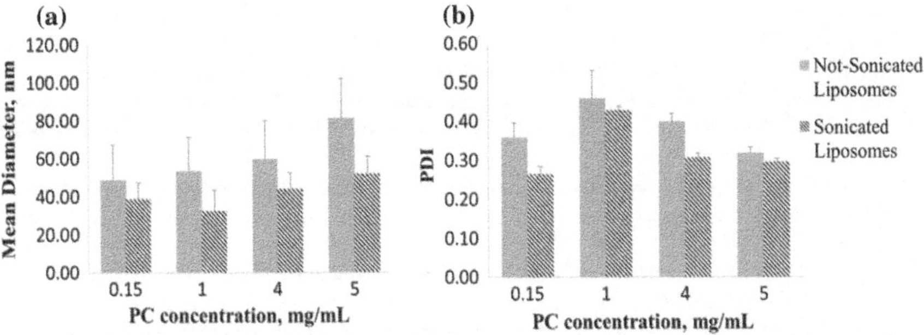

Fig. 4 **a** Sonicated and not sonicated liposomes diameter size at different PC concentrations in the hydroalcoholic solution. **b** Polydispersity Index (PDI) of sonicated and not sonicated liposomes at different PC concentrations in the hydroalcoholic solution. Results are expressed as average of three determinations and reported along with the standard deviation

For PDI values it seems that increasing the PC concentration the sample size distribution improves except for the 0.15 mg/ml concentration (Fig. 4b). The sonication in this case shows to have (except for the 0.15 mg/ml) a slight positive effect on the size distribution (values statistically significant, $p < 0.05$) confirming its suitability as homogenization treatment.

4 Conclusions

Nanoliposomes have been achieved by using a protocol based on the microfluidic principles. Effects of solutions flow rates and lipids concentrations on size and size distribution have been investigated. Increasing the ratio between the water volumetric flow rate to the lipids-ethanol volumetric flow rate the liposomes dimension decreases; at equal flow rates, when the lipids concentration increases also the liposomes size has been observed increasing. Ultrasonic energy was used to enhance homogenization of the hydroalcoholic bulk and, in turns to limit liposomes aggregation.

References

1. Attama, A.A., Momoh, M.A., Builders, P.F.: Lipid nanoparticulate drug delivery systems: a revolution in dosage form design and development, pp. 107–140. InTech, Croatia (2012)
2. Sawant, R.R., Torchilin, V.P.: Challenges in development of targeted liposomal therapeutics. The AAPS J. **14**(2), 303–315 (2012)
3. Reza Mozafari, M., et al.: Nanoliposomes and their applications in food nanotechnology. J. Liposome Res. **18**(4), 309–327 (2008)

4. Bochicchio, S., et al.: Vitamin delivery: carriers based on nanoliposomes produced via ultrasonic irradiation. LWT-Food Sci. Technol. **69**, 9–16 (2016)
5. Meure, L.A., Foster, N.R., Dehghani, F.: Conventional and dense gas techniques for the production of liposomes: a review. AAPS Pharmscitech **9**(3), 798–809 (2008)
6. Wagner, A., Vorauer-Uhl, K.: Liposome technology for industrial purposes. J Drug Deliv., 9 (2011)
7. Bangham, A., Horne, R.: Negative staining of phospholipids and their structural modification by surface-active agents as observed in the electron microscope. J. Mol. Bio. **8**(5), 660-IN10 (1964)
8. Pradhan, P., et al.: A facile microfluidic method for production of liposomes. Anticancer Res. **28**(2A), 943–947 (2008)
9. Barba, A., et al.: Ultrasonic energy in liposome production: process modelling and size calculation. Soft Matter **10**(15), 2574–2581 (2014)
10. Jahn, A., et al.: Controlled vesicle self-assembly in microfluidic channels with hydrodynamic focusing. J. Am. Chem. Soc. **126**(9), 2674–2675 (2004)

Production of Nanostructured Microspheres Biopolymer-Active Principle-Magnetic Nanoparticles by Supercritical Assisted Atomization

Renata Adami⑩, Mariarosa Scognamiglio⑩ and Ernesto Reverchon⑩

Abstract Supercritical Assisted Atomization (SAA) has been applied to the production of nanostructured microspheres ampicillin-chitosan-magnetic nanoparticles (AMP-CH-NMPs). Several ampicillin/chitosan (AMP/CH) ratios with a fixed content of NMPs were processed in acid water solutions, to produce microspheres with different size, drug content and amount of NMPs. To verify the successful formation of microparticles, drug content and nanoparticle dispersion, they were characterized by SEM (Scanning Electron Microscope), EDX (Energy Dispersive X-ray), TGA (ThermoGravimetric Analysis), HPLC (High Performance Liquid Chromatography), UV-vis obtaining information on morphology, particle size distribution, nanostructure, loading of active principle in the polymeric matrix and drug release rate. Spherical microparticles were obtained, with a maximum particle size of 2 μm and loading efficiencies up to 99%. The microspheres produced by SAA showed a controlled release of the drug over about 72 h.

Keywords Supercritical assisted atomization · Magnetic nanoparticles · Nanostructured microparticles · Controlled release · Targeted delivery · Ampicillin trihydrate

1 Introduction

Nanoparticles with Magnetic Properties (NMPs) can be used as contrast agent in magnetic resonance imaging (MRI) [1], as hyperthermia agents and as vectors for the targeted delivery of drugs; in this case they will be coprecipitated with the drug and driven to the target tissue [2]. NMPs have unique physical properties and ability to act at the cellular and molecular level of biological interactions [3].

R. Adami (✉) · M. Scognamiglio · E. Reverchon
Department of Industrial Engineering, University of Salerno,
Via Giovanni Paolo II 132, 84084 Fisciano, SA, Italy
e-mail: radami@unisa.it

© Springer International Publishing AG 2018
S. Piotto et al. (eds.), *Advances in Bionanomaterials*,
Lecture Notes in Bioengineering, DOI 10.1007/978-3-319-62027-5_2

Encapsulation of antibiotics in nanoparticles increases the maximum tolerated dose and therapeutic index of antibiotics when compared with the free drug [4].

In this work Supercritical Assisted Atomization (SAA) has been used for the production of nanostructured microspheres biopolymer-ampicillin-NMPs for biomedical applications. The aim was to entrap the drug and the NMPs in a polymer carrier that has the role to protect the active principle and to control its release. Moreover, the encapsulation in a polymeric coating makes the NMPs biocompatible and the polymer provides a steric barrier to prevent nanoparticle agglomeration [5]. Chitosan (CH) has been selected: it is a biopolymer of great interest as a carrier because of its natural origin; ampicillin tri-hydrate (AMP) has been chosen as model drug; Fe_3O_4 nanoparticles have been used as NMPs.

In literature there are some attempts at loading ampicillin in chitosan, in form of beads (850–1100 μm) or microgranules by coagulation of complex pastes followed by drying or in form of nanoparticles by ionic gelation method with the aid of sonication [6–8]. Nanoparticles of ampicillin trihydrate loaded in chitosan demonstrated superior antimicrobial activity with respect to plain nanoparticles and the reference, due probably to the synergistic effect of chitosan and the drug. In vitro release data showed an initial burst followed by slow sustained drug release [7].

The use of NMPs for treating infections is not well documented, on the other hand antibiotics are widely used in modern medicine [9]. Nanoparticle-based antimicrobial agents exhibit several advantages, including their inhibitory and/or killing effect on various microorganisms [10]. Hussein-Al-Ali et al. [11] prepared AMP-CH-Fe_3O_4 nanocomposite using a three step process: first NMPs were prepared through an iron salt coprecipitation method in an alkaline medium, followed by a chitosan coating step for the production of CH-coated NMPs, finally, the NMPs were loaded with AMP mixing AMP aqueous solution with CH-NMPs to form AMP-CH-NMP nanocomposite. The synthesized nanocomposites exhibited antibacterial and antifungal properties, as well as antimycobacterial effects.

To overcome the limitations of the traditional techniques, supercritical fluid based processes have been developed for several applications [12–14]. In particular, Supercritical Antisolvent (SAS), Supercritical Assisted Atomization (SAA), Supercritical Fluid Extraction from Emulsion (SEE) have been developed to produce micro and nanoparticles [15–25]. SAA has been successfully used for the production of co-precipitates, as nano-structured microparticles formed by a carrier in which the active principle is uniformly distributed [26–31] or as nanostructured polymeric microparticles loaded with nanoparticles [32, 33]. SAA is based on the formation of organic solvent + solid solutions in which SC-CO_2 is solubilized to form an expanded liquid of reduced viscosity and surface tension. These conditions produce an improved atomization, producing controlled micro and submicro droplets that, upon drying, form the corresponding polymer + drug coprecipitated microparticles [27, 28].

In previous works, the formation of polymer microparticles loaded with NMPs [32] and the production of AMP/CH microspheres [34] using SAA technique has been studied. In this work the formation of complex composite microparticles is proposed, starting from a suspension in which NMPs have been finely dispersed in a solution containing CH and AMP. It is the first time that a complex mixture of three compounds is processed by SAA. The aim is to obtain coprecipitates that can be applied in the targeted delivery and allow the sustained release of the drug when it is on the site of delivery. Several ratios AMP/CH are investigated, namely 1/2, 1/4 and 1/6 wt/wt and the effect on particle size distribution and AMP controlled release are studied.

2 Materials and Methods

Chitosan, medium molecular weight (CH), ampicillin trihydrate (AMP, Fig. 1) and Fe_3O_4 nanopowder (d_n < 50 nm, >98% trace metal basis) were supplied by Sigma-Aldrich (Milan, Italy). Water (HPLC grade) with a purity of 99.5% was supplied by Sigma-Aldrich (Milan, Italy). Carbon dioxide (CO_2; purity 99.9%) and Nitrogen (N_2; purity 99.9%) were purchased from SON (Naples, Italy).

In order to verify the stability of the suspension before the micronization process, solutions of acidic water at 1% v/v acetic acid and CH were prepared and Tween 80 was chosen as surfactant, because of its compatibility with water and with pharmaceutical processes. Solutions at concentrations of CH of 10 and 20 mg/mL and Tween 80 at different % wt/wt surfactant/solvent were sonicated in an ultrasound bath for 10 min. Then, increasing amounts of Fe_3O_4 nanoparticles (mean diameter 50 nm) were suspended in the solution and sonicated using a high-intensity ultrasonic probe at 50% amplitude for 1 min at pulse mode and their stability was monitored. SAA laboratory apparatus used is composed by two high-pressure pumps (mod. 305, Gilson, Villiers Le Bel, France) to deliver the water solution and the liquid CO_2 to the saturator. The saturator is a heated high pressure packed vessel (volume:25 cm^3) which assures a large contact surface between liquid solution and CO_2. The expanded liquid obtained in the saturator is sprayed through a nozzle into the precipitator (v = 3 dm^3) that operates at atmospheric pressure. A controlled heated flow of N_2 (about 1200 nL/h) is flown to the precipitator to enhance the evaporation of water from the droplets. A sintered filter at the bottom of the precipitator, with a porosity of 0.5 µm, allows the collection of the powder and the flowing through of the gases. SAA apparatus layout and further details can be found in previous papers [24].

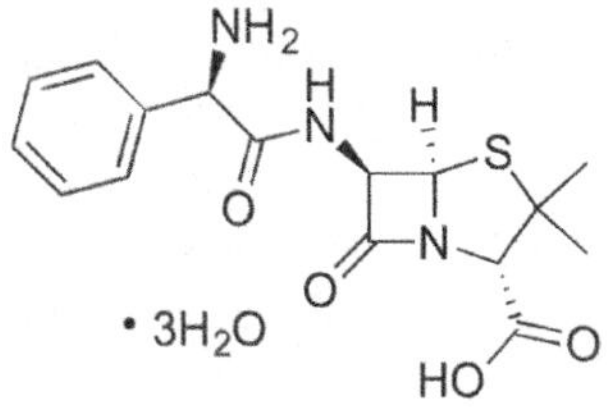

Fig. 1 Molecular structure of Ampicilin trihydrate

The morphology of produced particles has been analysed by a Field Emission Scanning Electron Microscope (FESEM, mod. LEO 1525, Carl Zeiss SMT AG, Oberkochen, Germany). Powders were dispersed on a carbon tab previously stuck to an aluminium stub (Agar Scientific, United Kingdom). Samples were coated with gold (layer thickness 150 Å) using a sputter coater (mod. 108 Å, Agar Scientific, Stansted, United Kingdom). Several SEM images were taken for each batch to verify the powder uniformity.

Particle size (PS) and particle size distribution (PSD) were measured from FESEM images using the Sigma Scan Pro Software (rel. 5.0, Jandel Scientific, Erkrath, Germany). Histograms representing PSDs in terms of volumetric cumulative were best fitted using Microcal Origin Software (rel. 8.5, Microcal Software, Inc., Northampton, MA).

SAA coprecipitates were also characterized by Energy-dispersive X-ray spectroscopy (EDX) microanalysis. Elemental analysis and element mapping were performed using the FESEM equipped with an energy dispersive X-ray spectroscopy (EDX, INCA Energy 350, Oxford Instruments, Witney, United Kingdom).

The loading of NMPs in the polymeric matrix was determined by thermogravimetric analysis (TGA) (SDT Q600, TGA/DSC, TA Instruments, USA), as the residue remaining after the total combustion of the organic component of the particles. The sample was placed in alumina crucibles and heated from 25 up to 600 °C at a rate of 10 °C/min, under a constant air flow of 100 Ncm^3/min).

Solid state analysis of the samples (XRPD = X-ray powder diffraction) was performed using an X-ray diffractometer (mod. D8 Discover, Bruker AXS, Inc., Madison, WI) with a Cu sealed tube source. Samples were placed in the holder and flattened with a glass slide to assure a good surface texture. The measuring conditions were as follows: Ni-filtered CuKα radiation, $\lambda = 1.54$ A, 2θ angle ranging from 2 to 90° with a scan rate of 3s/step and a step size of 0.02°.

Drug loading was performed dissolving Samples of AMP/CH powder under vigorous stirring in a solution water-acetic acid 2% v/v (pH 2.24) at 37 °C. The solution was stirred for 5 min and stored for 24 h at room temperature to obtain the complete release of the drug from the carrier. The obtained solution was filtered to eliminate the NMPs and CH residue and diluted with water to increase the pH and analysed by HPLC-UV/vis (Hewlett-Packard model G131–132, USA). The column used is a reverse phase C18 column (4.6 × 250 mm; 5 mm particle size; 80Å pore size; Hypersil BDS RP-C18); it was equilibrated at a flow rate of 0.5 mL/min with a mobile phase consisting of phosphate buffer pH 5.0 and acetonitrile (ratio 70:30 v/v). AMP was monitored at 225 nm with a retention time of 5–6 min. Loading efficiency was calculated as the ratio of the drug content of the produced particles over the drug loaded at the beginning.

The drug release rate was analysed by UV/vis spectrophotometer (model Cary 50, Varian, Palo Alto, CA) using a physiologic solution (PS) at pH 6.8 as dissolution medium. 40 mg of powder was charged into a tea bag filter and immerged in 200 mL of PS continuously stirred at 200 rpm and 37 °C. Absorbance at wavelength 264 nm was monitored as: every 1 min for 1 h, every 10 min for 1 h, every

30 min for 8 h, every 60 min for 72 h. Then, the absorbance values were converted into AMP concentration, using a calibration curve.

3   Results and Discussion

The system processed by SAA was a suspension of NMPs in an aqueous solution of CH and AMP; therefore, preliminary studies were performed on the suspension in terms of stability of NMPs dispersion and degradation of the drug.

3.1   Stability of Suspensions

For a successful SAA precipitation, it is relevant to have a stable suspension, that does not split before precipitation. For this reason, in a previous work, a study on the suspension stability has been performed [32]. The suspensions were considered useful for SAA processing if they were stable for at least 30 min. Such suspensions were obtained at a chitosan concentration of 10 mg/mL plus 2% Tween 80 (w/w surfactant/solvent) with a maximum loading of 20% of Fe_3O_4 nanoparticles (w/w NMPs/chitosan) and at a chitosan concentration of 20 mg/mL plus 7.5% Tween 80 (w/w surfactant/solvent) with a maximum load of 40% NMPs (w/w NMPs/chitosan).

3.2   Micronization Experiments

The micronization of the system AMP/CH by SAA has been studied in a previous work [34] and several operating conditions have been investigated. Based on the previous results [32], the operating conditions for the experiments performed in this work have been chosen as reported in Table 1. For all the experiments a ratio between CO_2 and liquid solution (gas to liquid ratio = GLR) of 1.8 have been used.

Table 1 Operating conditions of SAA experiments varying the global concentration, R = AMPt/CH, conc = solute concentration, P_m = mixer pressure, T_m = mixer temperature, P_p = precipitator pressure, T_p = precipitator temperature, Q = flow rate

Conc. CH (mg/mL)	Conc. AMP (mg/mL)	R (wt/wt)	T_m (°C)	P_m (bar)	T_p (°C)	P_p (bar)	$Q_{suspension}$ (mL/min)	Q_{CO2} (gr/min)
10	5	1/2	90	90	113	1.6	4.48	8.06
10	2.5	1/4	91	99	112	1.8	4.62	8.31
10	1.67	1/6	80	109	113	1.1	4.41	7.94

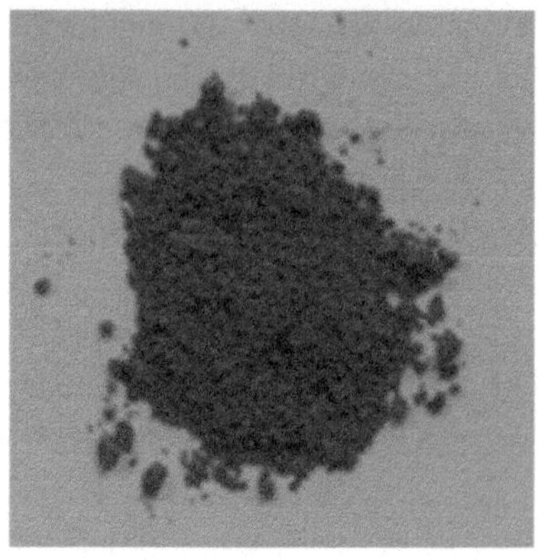

Fig. 2 Example of powder obtained by SAA of coprecipitate AMP/CH containing 40% di Fe_3O_4 (wt/wt_{CH})

The following SAA process parameters have been selected: static mixer temperature and pressure at 90 °C and 100 bar, respectively; precipitator temperature set at 110 °C and a concentration of CH of 10 mg/mL. 40% wt of NMPs have been suspended in the starting solution and AMP/CH ratios of 1/2, 1/4, 1/6 have been studied.

After each experiment, a dry powder with a brownish colour (Fig. 2) was collected on the filter of the precipitator, giving a first evidence of the presence of NMPs uniformly distributed in the particles.

In all the experiments the particles obtained showed a spherical morphology, as in the Fig. 3, that reports, as an example, SEM photomicrographs of coprecipitates at AMP/CH ratios 1/2, 1/4, 1/6 (wt/wt). By increasing CH content a characteristic wrinkling of the particle surface can be noted.

In Fig. 4 particle size distributions in terms of cumulative volumetric percentages are reported, showing that the mean diameter ranges between 0.5 and 1 μm and the maximum observed particle size is about 2 μm. The results did not show a clear influence of drug/polymer ratio on the particle size and distribution in the case of R = 1/2 and R = 1/4, and there is a slight increase of particle size in case of R = 1/6.

The obtained microspheres have been characterized by several analytical techniques to verify the efficiency of SAA micronization. XRPD analyses of the coprecipitates showed the presence of the characteristic halo typical of an amorphous material and Fe_3O_4 crystal peaks (Fig. 5). Therefore, the microspheres are amorphous, the drug is intimately mixed with the polymer and, during the precipitation, the NMPs do not modify their characteristic crystalline structure: they are simply entrapped in the polymeric matrix.

EDX microanalysis allowed to identify the elemental composition of the particles in a SEM photomicrograph, showing the presence of the elements characteristic of the three compounds (Fig. 6). Oxygen (white in Fig. 6) is present in all the compounds, Sulphur (yellow in Fig. 6) is present only in the AMP molecule, as can be observed in Fig. 1 reporting the molecular structure, and Iron (cyan in Fig. 6) is present only in Fe_3O_4. Figure 6 confirms that the three elements are present in the

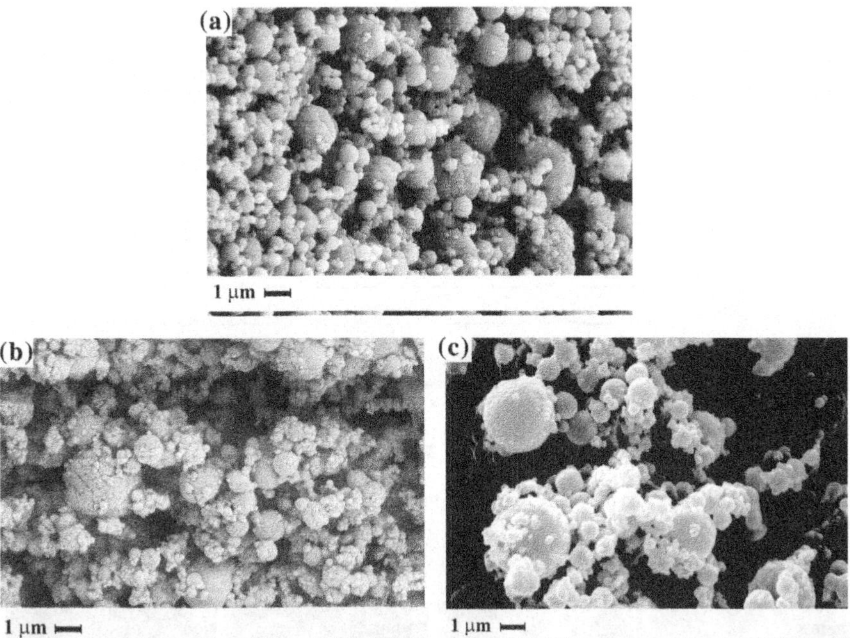

Fig. 3 SEM photomicrographs of SAA coprecipitate containing 40% di Fe_3O_4 at different R = AMP/CH: **a** R = 1/2, **b** R = 1/4, **c** R = 1/6

Fig. 4 PSD of AMP/CH loaded with 40% NMPs micronized by SAA at different R (AMP/CH ratio, wt/wt), in terms of cumulative volumetric percentage. *Dashed line* R = 1/6, *Continuous line* R = 1/4, *Dotted line* R = 1/2

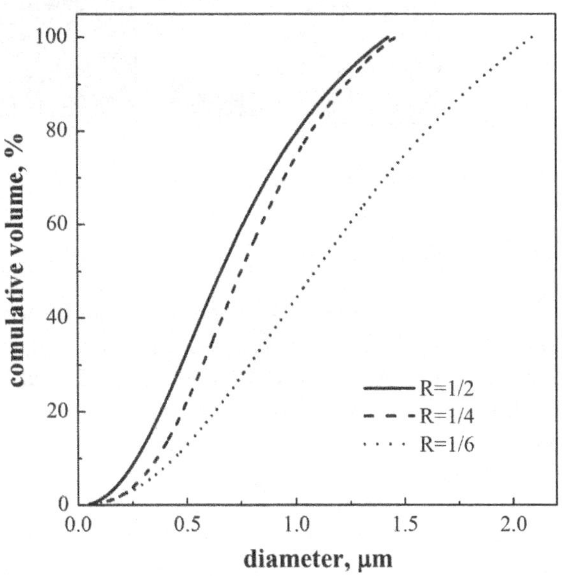

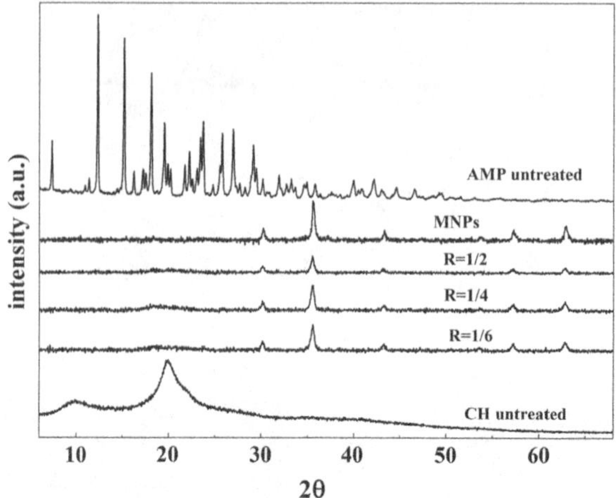

Fig. 5 XRPD spectra of powders: comparison among untreated materials and microspheres produced by SAA

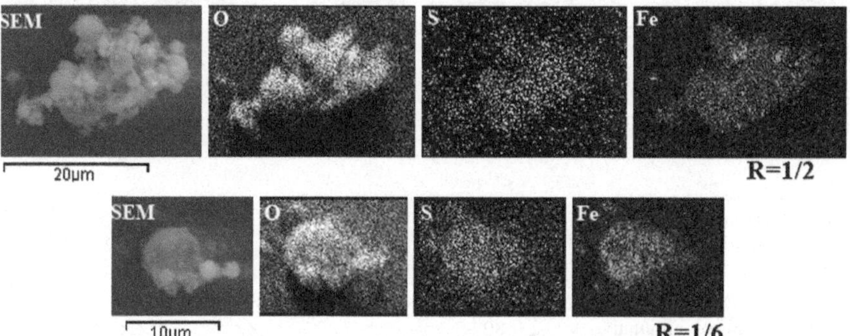

Fig. 6 EDX microanalysis of SAA coprecipitated microparticles loaded with 40% Fe_3O_4. Particulars of: SEM image of the analysed areas; Oxygen maps (*white*); Sulphur maps (*yellow*); Iron maps (*cyan*). R = AMPt/CH ratio, wt/wt

coprecipitates; therefore, AMP and NMPs are loaded in the CH particles; furthermore, the key chemical elements are uniformly distributed in the microparticles, confirming that AMP and CH are intimately mixed in each particle at nanometric level. Furthermore, it is shown that the NMPs are also uniformly distributed in the microparticles, mainly as a result of the stability of the suspension during the micronization process.

To obtain a quantitative information about the NMPs present in the microparticles, TGA analysis has been performed on the coprecipitates obtained by SAA producing the combustion of the organic material. The organic components CH and AMP decompose by combustion, leaving the inorganic residue Fe_3O_4. The quantity

Table 2 Loading efficiency of AMP and NMPs in SAA coprecipitates. Fe_3O_4 starting concentration 40%

Conc. CH (mg/mL)	R = AMP/CH wt/wt	Fe_3O_4 loading efficiency (%)	Fe_3O_4 loaded (wt%)	AMP loading efficiency (%)	AMP loaded (wt%)
10	1/2	37	14.81	63	20.95
10	1/4	31	12.47	66	13.24
10	1/6	31	12.25	99	14.12

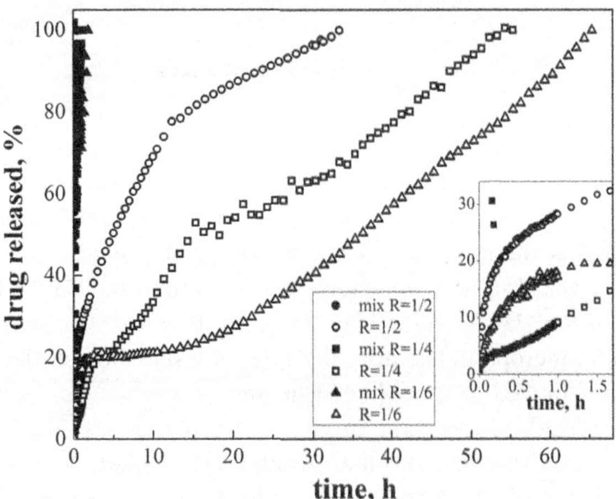

Fig. 7 Release profiles from microspheres produced by SAA and physical mixtures at different drug loadings; R = AMP/CH ratio wt/wt, mix = physical mixture

of NMPs loaded in the microparticles varies between 31 and 37%, as reported in Table 2.

HPLC analyses showed that a loading efficiency of the active principle in the range of 63–99%, as reported in Table 2. AMP loading efficiency increases with the decrease of the AMP/CH ratio; this is because the increase of the relative amount of CH improves the entrapment of the active principle.

Drug release studies have been performed using a filter bag in physiologic solution at pH 6.8 and 37 °C. CH is very soluble at very low pH values, on the contrary, at pH close to neutral, characteristic of human tissue, it has a very low solubility and can be used for delayed release. The controlled drug release studies were elaborated considering the measured drug loading: the maximum percentage of drug released from the microspheres was normalized on the drug effectively loaded in the coprecipitate. The release form microspheres was compared with the release form physical mixtures of AMP, CH and NMPs at the same ratios than in the microspheres.

As can be noted in Fig. 7, physical mixtures release the drug within 55 min, mainly because AMP is not entrapped in the CH and dissolves in the aqueous environment. The release of AMP from CH microspheres is slower compared to the physical mixtures, and the release rate is controlled by the AMP/CH ratio. It can be

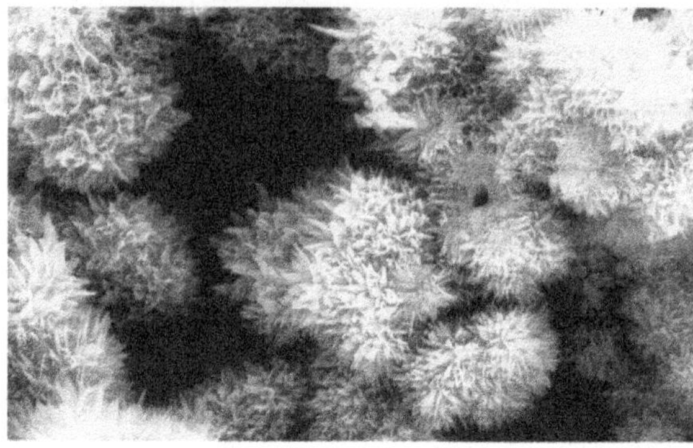

Fig. 8 SEM photomicrographs of SAA coprecipitate containing 40% di Fe_3O_4 at R = 1/4, after the AMP release

observed for release of AMP at all the ratios studied a fast release in the first 150 min, followed by a slow release, that is different for the AMP/CH ratios: the higher the amount of CH, the slower the release.

The drug is completely released after about 33 h for R = 1/2, after about 55 h for R = 1/4 and after about 65 h for R = 1/6. The observation of the morphology of the microparticles, after the release tests, showed that the particles were dissolved and moved to a gel-like structure.

SEM observation of the microspheres, showed that the presence of NMPs affected the release: the morphology of particles showed that the NMPs opened several holes on their surface during their release (Fig. 8).

4 Conclusions

Composite microparticles chitosan-ampicillin loaded with Fe_3O_4 nanoparticles were successfully produced by SAA and can be used for the production of pharmaceutical formulation for the targeted release of the active principle in specific sites. The spherical morphology and controlled PSD with a maximum diameter of about 2 μm allow to use the microspheres for different biomedical applications. High loading efficiency of AMP in CH microspheres can be obtained. The release of the active principle can be controlled by the AMP/CH ratio and there is no interference of NMPs to the release of AMP from the formulation. A starting burst effect followed by a sustained release allow the formulation to be effective up to 72 h for a AMP/CH 1/6.

Acknowledgements The authors gratefully acknowledge Dr. Valentina Gregori and Dr. Alessia Di Capua for the help in performing the experiments. The MiUR (Ministero dell'istruzione, dell'Università e della Ricerca) is acknowledged for the financial support.

References

1. Felinto, M.C.F.C., Camilo, R.L., Diegues, T.G.: Magnetic nanoparticles and their application in biomedicine. In: International Nuclear Atlantic Conference—INAC 2007, Satnos, SP, Brazil (2007)
2. Neuberger, T., Schopf, B., Hofmann, H., Hofmann, M., von Rechenberg, B.: Superparamagnetic nanoparticles for biomedical applications: possibilities and limitations of a new drug delivery system. J. Magn. Magn. Mater. **293**, 483–496 (2005)
3. Chomoucka, J., Drbohlavova, J., Huska, D., Adam, V., Kizek, R., Hubalek, J.: Magnetic nanoparticles and targeted drug delivering. Pharmacol. Res. **62**, 144–149 (2010)
4. Pinto-Alphandary, H., Andremont, A., Couvreur, P.: Targeted delivery of antibiotics using liposomes and nanoparticles: research and applications. Int. J. Antimicrob. Ag. **13**, 155–168 (2000)
5. Faraji, A.H., Wipf, P.: Nanoparticles in cellular drug delivery. Bioorgan. Med. Chem. **17**, 2950–2962 (2009)
6. Chandy, T., Sharma, C.P.: Chitosan matrix for oral sustained delivery of ampicillin. Biomaterials **14**, 939–944 (1993)
7. Saha, P., Goyal, A.K., Rath, G.: Formulation and evaluation of chitosan-based ampicillin trihydrate nanoparticles. Tropical J. Pharm. Res. **9**, 483–488 (2010)
8. Changerath, R., Nalr, P.D., Mathew, S., Fteghunadhan, C.P.: Nalr: Poly(methyl methacrylate)-grafted chitosan microspheres for controlled release of ampicillin. J. Biomed. Mater. Res.—Part B Appl. Biomater. **89**, 65–76 (2009)
9. Ball, A.P., Bartlett, J.G., Craig, W.A., Drusano, G.L., Felmingham, D., Garau, J.A., Klugman, K.P., Low, D.E., Mandell, L.A., Rubinstein, E., Tillotson, G.S.: Future trends in antimicrobial chemotherapy: expert opinion on the 43rd ICAAC. J. Chemother. **16**, 419–436 (2004)
10. Huh, A.J., Kwon, Y.J.: "Nanoantibiotics": a new paradigm for treating infectious diseases using nanomaterials in the antibiotics resistant era. J. Control Release **156**, 128–145 (2011)
11. Hussein-Al-Ali, S.H., El Zowalaty, M.E., Hussein, M.Z., Geilich, B.M., Webster, T.J.: Synthesis, characterization, and antimicrobial activity of an ampicillin-conjugated magnetic nanoantibiotic for medical applications. Int. J. Nanomed. **9**, 3801–3814 (2014)
12. Baldino, L., Cardea, S., Reverchon, E.: Production of antimicrobial membranes loaded with potassium sorbate using a supercritical phase separation process. Innov. Food Sci. Emerg. **34**, 77–85 (2016)
13. Reverchon, E., Adami, R.: Nanomaterials and supercritical fluids. J. Supercrit. Fluid **37**, 1–22 (2006)
14. Reverchon, E., Sesti Osseo, L., Gorgoglione, D.: Supercritical CO_2 extraction of basil oil: characterization of products and process modeling. J. Supercrit. Fluids **7**, 185–190 (1994)
15. Rossmann, M., Braeuer, A., Schluecker, E.: Supercritical antisolvent micronization of PVP and ibuprofen sodium towards tailored solid dispersions. J. Supercrit. Fluids **89**, 16–27 (2014)
16. De Paz, E., Martín, Á., Every, H., Cocero, M.J.: Production of water-soluble quercetin formulations by antisolvent precipitation and supercritical drying. J. Supercrit. Fluid **104**, 281–290 (2015)
17. Kurniawansyah, F., Mammucari, R., Foster, N.R.: Inhalable curcumin formulations by supercritical technology. Powder Technol. **284**, 289–298 (2015)
18. Santiago, L.M., Masmoudi, Y., Tarancón, A., Djerafi, R., Bagán, H., García, J.F., Badens, E.: Polystyrene based sub-micron scintillating particles produced by supercritical anti-solvent precipitation. J. Supercrit. Fluid **103**, 18–27 (2015)
19. Nerome, H., Machmudah, S., Wahyudiono, R., Fukuzato, T., Higashiura, H., Kanda, M.Goto: Effect of solvent on nanoparticle production of β-Carotene by a supercritical antisolvent process. Chem. Eng. Technol. **39**, 1771–1777 (2016)
20. Shen, Y.B., Du, Z., Wang, Q., Guan, Y.X., Yao, S.J.: Preparation of chitosan microparticles with diverse molecular weights using supercritical fluid assisted atomization introduced by hydrodynamic cavitation mixer. Powder Technol. **254**, 416–424 (2014)

21. Wu, H.T., Yang, M.W., Huang, S.C.: Sub-micrometric polymer particles formation by a supercritical assisted-atomization process. J. Taiwan Inst. Chem. E. **45**, 1992–2001 (2014)
22. Prosapio, V., De Marco, I., Reverchon, E.: PVP/corticosteroid microspheres produced by supercritical antisolvent coprecipitation. Chem. Eng. J. **292**, 264–275 (2016)
23. Della Porta, G., Campardelli, R., Cricchio, V., Oliva, F., Maffulli, N., Reverchon, E.: Injectable PLGA/Hydroxyapatite/Chitosan microcapsules produced by supercritical emulsion extraction technology: an in vitro study on Teriparatide/Gentamicin controlled release. J. Pharm. Sci.-Us. **105**, 2164–2172 (2016)
24. Adami, R., Liparoti, S., Reverchon, E.: A new supercritical assisted atomization configuration, for the micronization of thermolabile compounds. Chem. Eng. J. **173**, 55–61 (2011)
25. Wu, H.T., Huang, S.C., Yang, C.P., Chien, L.J.: Precipitation parameters and the cytotoxicity of chitosan hydrochloride microparticles production by supercritical assisted atomization. J. Supercrit. Fluid **102**, 123–132 (2015)
26. Shen, Y.B., Guan, Y.X., Yao, S.J.: Supercritical fluid assisted production of micrometric powders of the labile trypsin and chitosan/trypsin composite microparticles. Int. J. Pharm. **489**, 226–236 (2015)
27. Labuschagne, P.W., Adami, R., Liparoti, S., Naidoo, S., Swai, H., Reverchon, E.: Preparation of rifampicin/poly(D, L-lactice) nanoparticles for sustained release by supercritical assisted atomization technique. J. Supercrit. Fluid **95**, 106–117 (2014)
28. Liparoti, S., Adami, R., Caputo, G., Reverchon, E.: Supercritical assisted atomization: polyvinylpyrrolidone as carrier for drugs with poor solubility in water. J. Chem.-Ny. (2013)
29. Adami, R., Liparoti, S., Della Porta, G., Del Gaudio, P., Reverchon, E.: Lincomycin hydrochloride loaded albumin microspheres for controlled drug release, produced by supercritical assisted atomization. J. Supercrit. Fluid **119**, 203–210 (2017)
30. Shen, Y.B., Du, Z., Tang, C., Guan, Y.X., Yao, S.J.: Formulation of insulin-loaded N-trimethyl chitosan microparticles with improved efficacy for inhalation by supercritical fluid assisted atomization. Int. J. Pharm. **505**, 223–233 (2016)
31. Wu, H.T., Yang, C.P., Huang, S.C.: Dissolution enhancement of indomethacin-chitosan hydrochloride composite particles produced using supercritical assisted atomization. J. Taiwan Inst. Chem. E. **67**, 98–105 (2016)
32. Adami, R., Reverchon, E.: Composite polymer-Fe_3O_4 microparticles for biomedical applications, produced by supercritical assisted atomization. Powder Technol. **218**, 102–108 (2012)
33. Reverchon, E., Adami, R.: Supercritical assisted atomization to produce nanostructured chitosan-hydroxyapatite microparticles for biomedical application. Powder Technol. **246**, 441–447 (2013)
34. Reverchon, E., Antonacci, A.: Drug-polymer microparticles produced by supercritical assisted atomization. Biotechnol. Bioeng. **97**, 1626–1637 (2007)

Encapsulation of Hydrophilic and Lipophilic Compounds in Nanosomes Produced with a Supercritical Based Process

Paolo Trucillo, Roberta Campardelli and Ernesto Reverchon

Abstract Liposomes are created when phospholipids self-assemble in an aqueous medium creating spherical closed structures. These vesicles can be loaded with hydrophilic active principles (AP) into the aqueous inner core or with lipophilic compounds in the lipidic double layer. In this work a new supercritical based process for the one-step continuous production of nanosomes is proposed for the encapsulation of hydrophilic and lipophilic compounds. This process is called Supercritical Assisted Liposome Formation (SuperLip). The innovation of this process consists in the inversion of the traditional phases of production of liposomes: water droplets are created by a spray atomization in a high pressure vessel, and then a double layer of phospholipids fast surrounds them. A systematic study on liposome size, morphology, encapsulation efficiency has been performed for several different hydrophilic AP (ampicillin, ofloxacin, bovine serum albumin, fluorescein, eugenol and theophylline). Some operative parameters were also optimized to achieve the production of nanometric liposomes with high encapsulation efficiencies. Operating in this way nanometric and monodispersed liposome suspensions were produced with EE up to 99%. To complete the study, other lipidic compounds were entrapped in the double lipidic layer, obtaining high entrapment efficiencies (TE), also in this case, up to 84.9%.

Keywords Liposomes · Drug carriers · Supercritical fluids

Abbreviations

AP	Active Principles
EE	Encapsulation Efficiency
FV	Formation Vessel
S	Saturator
SS	Separation Step

P. Trucillo · R. Campardelli (✉) · E. Reverchon
Department of Industrial Engineering, University of Salerno, Via Giovanni Paolo II, 132, 84084 Fisciano, SA, Italy
e-mail: rcampardelli@unisa.it

© Springer International Publishing AG 2018
S. Piotto et al. (eds.), *Advances in Bionanomaterials*,
Lecture Notes in Bioengineering, DOI 10.1007/978-3-319-62027-5_3

DLS   Dynamic Light Scattering
MD   Mean Diameter
SD   Standard Deviation
TE   Entrapment Efficiency
SEM   Scanning Electron Microscope
TEM   Transmission Electron Microscopy
Exp   Experiment
PSD   Particle Size Distributions

1   Introduction

Generally, the use of drug carriers [1] enhances the bioavailability of drugs [2], while they are reaching the target cell or tissue. For this reason, the entrapment of a molecule inside a biocompatible shell or a vesicle gives the possibility to protect the drug from acidity, toxicity, light and heat [3]. In fact, external stimuli can be dangerous for the preservation of drugs. Moreover, without a drug carrier, it is difficult to achieve a successful targeted drug delivery.

For the entrapment of active principles (AP), nanotechnology has developed many kind of drug carriers. For example, nanospheres, nanocapsules and nanosomes were designed and produced [4, 5].

Among these, liposomes are lipidic vesicles characterized by one or more double lipidic external layers and a water inner core [6]. Lipophilic drugs can be entrapped inside the lipidic layer whereas the hydrophilic AP can be encapsulated in the inner core. Due to their similarity with living cell membranes, liposomes are used as non-toxic and biocompatible drug carriers for proteins [7], vitamins [8], antibiotics [9], vaccines [10], markers [11], anticancer [12], enzymes [13] and DNA and RNA for gene therapy [14–19]. Indeed, the favorable interactions with in vivo systems increased the interest in the optimizations of the vesicles composition. Liposomes were considered as a formidable and versatile model for cell membrane barrier that can be administrated with oral, gastrointestinal, intra-venous or via skin penetration delivery [20]. Drug encapsulated into liposomes generally showed a higher shelf-life with higher circulating time, reduced toxicity and the fundamental skill to reach the target cell with minor lack of its content. Furthermore, liposomes have a better cargo potential for drugs, compared to polymer particles.

Conventional processes were developed to produce liposomes loaded with different kind of drugs; for example, thin layer hydration is the simplest and the most used method since the first liposomes suspension was prepared [21]. But, conventional methods often suffered of many drawbacks, like batch layouts, production of giant micrometric vesicles, difficult reproduction of particle size distributions and low encapsulation efficiency (EE) of hydrophilic compounds. As a consequence of this, micrometric vesicles are created, with difficulties in the elimination of the organic solvent. Moreover, to increase the encapsulation efficiency, a further post-processing step of sonication has been generally performed. Furthermore, to reduce the dispersion of the samples, an extrusion step could be added to the vesicles formation.

Supercritical fluids have been successfully used to produce also polymeric and biocompatible nanoparticles for drug delivery applications [22–26].

The application of supercrtitical fluids technology has been proposed also to overcome the problems linked to conventional liposome preparation processes [27], Nevertheless, many supercritical processes developed to produce liposomes still present some drawbacks as low EE of hydrophilic AP. Also, post-processing is still required to obtain nanometric-sized liposomes.

Recently, a novel supercritical fluid assisted process has been developed for the production of liposomes [28]. This technique is called SuperLip (Supercritical Assisted Liposome formation) and exploits CO_2 peculiar properties of high diffusion coefficient and low viscosity in supercritical conditions. In fact, the fast diffusion permits the phospholipids to create one or more layer in a very fast manner compared to other processes. Indeed, in this process the first step is the production of the water droplets and then phospholipids fast surround them producing liposomes. In this way, small unilamellar vesicles are formed [29]. Furthermore, SuperLip is a continuous process, that can be scaled up easily and that can work for 24 h a day.

2 Materials

Among hydrophilic compounds, Fluorescein (MW 332.31 g/mol, >99.5 purity), Bovine Serum Albumin (MW 66463 g/mol, >99% purity), Ampicillin (MW 349.41 g/mol, 99% purity), Ofloxacin (MW 361.37, > 99.5%purity), Theophylline (MW 180.16, >99% purity), Eugenol (MW 164.20, purity >99%) were bought from Sigma Aldrich, Milan, Italy.

Among amphoteric and lipophilic compounds, alpha-Lipoic Acid (MW 203.66 g/mol, >99% purity), cholesterol (MW 386.67, >99% purity), L-α-phosphatidylcholine from egg (~60% purity), phosphatidylethanolamine (>97% purity) was supplied by Sigma Aldrich, Milan, Italy. All the compounds were used as received.

Among the solvents used to perform SuperLip process, ethanol was purchased by Sigma Aldrich, Milan, Italy (>99.8% purity). Carbon dioxide was provided by SON, Naples, Italy (>99.4% purity). Distilled water was self-produced in our laboratories.

3 Methods

3.1 Liposomes Dimensions

Liposome suspensions were characterized using Dynamic Light Scattering (DLS) technique. Using the instrument Mod. Zetasizer Nano S (Worcestershire, UK), liposomes mean diameters (MD) were measured, as well as polydispersion index (PDI) and standard deviation (SD) of the vesicles. DLS instrument works at 25 °C and is equipped with a 5.0 mW He–Ne laser operating at 633 nm with a scattering angle of 173°. Every sample was measured 3 times in order to calculate the mean values of the measurements.

3.2 Liposomes Morphology

Transmission electron microscopy, TEM, (JEOL 1400, 100 kV accelerating voltage) was used with negative staining as a further confirmation of liposomes morphology and mean diameter. For samples preparation, a droplet of liposome suspension was placed on a copper grid allowed to sit for 60 s. Then, the droplet was dried with filter paper. A droplet of staining agent was subsequently placed on top of the grid and left reacting for 30 s, the excess was then removed with a filter paper.

3.3 Encapsulation Efficiency

The determination of the encapsulation efficiency (EE) of liposomes was measured with the most used supernatant method [30] Triplicates of the analysis were performed and the results reported are the mean values of the EE. First, liposomes samples were ultra-centrifuged at 13,000 rpm for 50 min at −4 °C. Then, the supernatant was kindly removed from the top. The amount of drug in the supernatant was measured using a Micro-volume UV-Vis spectrophotometer (BioSpec-nano, Shimadzu Scientific Instruments, Columbia, USA) at various wave lengths depending on the AP. In details, for hydrophilic compounds encapsulated into liposomes, the wavelength for fluorescein UV- Vis assay was 515 nm, for bovine serum albumin 280 nm, for ampicillin 225 nm, for ofloxacin 290 nm, for theophylline 275 nm and for eugenol 280 nm. Among lipophilic or amphoteric compounds, the UV-Vis assay was 280 nm for eugenol and 325 nm and α-lipoic acid. The Encapsulation Efficiency was calculated as the complement to 100 of the percentage of drug present in the supernatant, as expressed mathematically in the Eq. (1).

$$EE[\%] = 100 * \left(1 - \frac{ppm_{supern}}{ppm_{loaded}}\right) \tag{1}$$

Instead, the amount of cholesterol entrapped in the lipidic double layer was measured with Gas Chromatography (GC) assay (GC-FID, mod. 6890 Agilent Series, Agilent Technologies Inc), according to the method reported in the literature [31]. A defined volume of liposome suspension was centrifuged at 13,000 rpm for 50 min at −4 °C, then the pellet was re-suspended in 3 mL of hexane and kindly agitated for 30 min. Finally 2 µL of this solution were used for GC analysis.

3.4 Statistical Analysis

The Particle Size Distributions (PSDs) of liposomes samples were all expressed in terms of mean value ± standard deviation (MV ± SD),determined using one-way analysis of variance (ANOVA) and Student's t-test, with a 0.05 level of probability considered as significant.

4 Apparatus

In this process, lipids and eventual lipophilic compounds were dissolved in ethanol. $SC\text{-}CO_2$ and ethanol solution were fed separately in a thermally heated saturator where an expanded liquid (EL) was obtained. The saturator worked at the pressure of 100 bar. Then, the EL was fed to a high pressure stainless steel Formation Vessel (FV) at the same pressure. A third feed line was used to add the water solution in the formation vessel. This solution was atomized to generate water droplets containing the hydrophilic drug. The atomization was performed through a nozzle of the diameter of 80 μm. Carbon Dioxide was pumped using an Ecoflow pump (mod. LDC-M-2, Lewa, Germany), whose head was continuously refrigerated by a thermostatic bath (mod. FL 300, Lauda, Germany), working with a 70:30 (vol/vol) ethylene glycol-water solution. Instead, water and ethanol solutions were pumped separately using high pressure precision pumps (Model 305, Gilson, France). An exit at the top of the FV was provided to eliminate the expanded liquid without contaminate the liposomes suspension, that was recovered from the bottom, instead. A fundamental decompression step was provided to separate CO_2 and ethanol, using a stainless-steel separator that operated at 10 bar and 25 °C. SuperLip lab-scale plant scheme is reported in our previous work [28, 32].

The flow rate of ethanol solution was set to 3.5 mL/min., and CO_2 was fed to the saturator at 6.5 g/min, together with ethanol solution. Mass based Gas to Liquid Ratio (GLR) was equal to 2.4. Working temperature was set to 40 °C. Water flow rate was fixed to 10 mL/min for all the experiments.

5 Results

5.1 Empty Liposomes

The first experiments were performed for the production of empty liposomes, to verify the possibility of producing nanometric and stable vesicles with this process. For these experiments, 500 mg of phosphatidylcholine (PC) were dissolved in 100 mL ethanol to form a lipidic solution. The mixture Ethanol + PC was pumped at the flow rate of 3.5 mL/min. The flow rate of carbon dioxide was set to

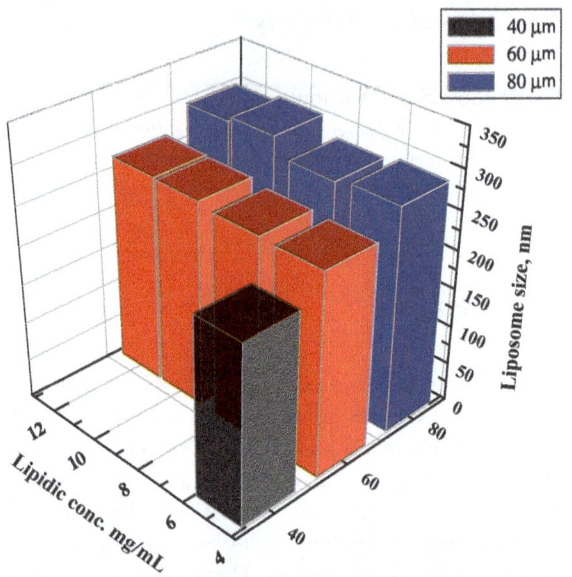

Fig. 1 Liposomes mean diameters as a function of lipid concentration and water flow nozzle diameter

6.5 g/min, with a GLR of about 2.4, mass based. The pressure was fixed to 100 bar and the temperature to 40 °C. The effect of the variation of two parameters was investigated: lipidic concentration and the diameter of the water atomization nozzle. Results are summarized in Fig. 1.

As reported in Fig. 1, the size of the liposomes produced are in the range between 200 and 300 nm. The mean diameters decreased when the diameters of the nozzle used for water atomization decreased. This is explained by the fact that the jet break up brings to the formation of smaller droplets when a smaller injector was used. Instead, the lipid concentration did not affect significantly liposomes mean diameters. A maximum concentration of 12 mg/mL was reached for the successful production of liposomes for 60 and 80 μm diameters of the nozzle. Above this limit, the nozzle risks to be blocked. This limit decreases to 5 mg/mL for 40 μm water injector diameter. No simultaneous effect of both parameters variation was observed.

5.2 Hydrophilic Compounds Encapsulation

Once established that SuperLip is able to produce liposomes of nanometric dimensions with a good control of particle size distribution, a systematic study on liposome size, polydispersion and encapsulation efficiency has been performed for the encapsulation of several compounds belonging to different families. Some operative parameters were also optimized to achieve the production of nanometric liposomes with high encapsulation efficiencies.

Table 1 Theoretical loadings, mean diameters (MD) and encapsulation efficiencies (EE) of many compounds entrapped in the inner core with SuperLip

Compound	Theoretical loading (%, w/w)	MD (nm ± SD)	EE (%)
Empty	–	290 ± 60	–
Fluorescein	3	277 ± 55	90.0
Bovine Serum Albumin	60	245 ± 73	93.9
Ampicillin	3	313 ± 69	99.0
Ofloxacin	1	256 ± 71	86.2
Theophylline	1	136 ± 86	98.0
Eugenol	20	131 ± 48	92.5

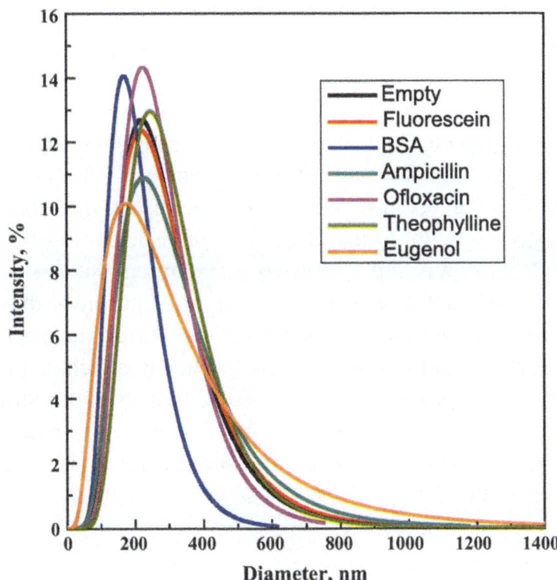

Fig. 2 Comparison of particle size distributions of the vesicles loaded with hydrophilic drugs

Operating in this way, nanometric and monodispersed liposome suspensions were produced with high encapsulation efficiencies, as reported in Table 1 and Fig. 2.

As reported in Table 1 and Fig. 2, six different compounds were encapsulated inside the water core of the liposomes. It was possible to achieve the production of liposomes with a mean diameter included between 131 ± 48 nm and 313 ± 69 nm. This means that the pressure and the temperature chosen for the process contributed to the production of water droplets smaller than 300 nm. Indeed, mean diameters are not affected by the kind of compound encapsulated. The high diffusion coefficient of carbon dioxide in supercritical conditions gave also the possibility to fast surround water droplets, avoiding a lack of the drug. In fact, the encapsulation efficiencies were high and included between 86.2 and 99.0%.

Table 2 Theoretical loadings, mean diameters (MD) and entrapment efficiencies (TE) of many compounds entrapped in the lipidic layer with SuperLip

Compound	Theoretical loading (%, w/w)	MD (nm ± SD)	TE (%)
Cholesterol	2.5	171 ± 119	81.6
PE	2.5	197 ± 40	80.5
Eugenol	30	229 ± 96	84.9
ALA	30	109 ± 49	63.2

5.3 Lipophilic Compounds Encapsulation

To complete the study, cholesterol and phosphatidylethanolamine were included as additives in the lipidic PC layer. To further confirm that the described process is able to encapsulate compounds successfully, amphoteric compounds as eugenol and ALA were entrapped in the lipidic layer compartment of liposomes. Mean diameters and encapsulation efficiencies are reported in Table 2.

As reported in Table 2, four different compounds were dissolved in the lipidic solution together with phosphatidylcholine for the entrapment inside the double layer of liposomes. Also in this case, the pressure and temperature conditions chosen brought to the production of vesicles whose diameters is included between 109 ± 49 nm and 229 ± 96 nm. The entrapment of the drugs inside the lipidic layer did not affect the mean size of liposomes that is possible to produce with SuperLip. In the best working conditions reported in Table 2, the entrapment Efficiency (TE) in the lipid compartment is included between a minimum of 63.2% and a maximum of 84.9%. Compared to hydrophilic entrapment efficiency, the average value of lipophilic TE is smaller than hydrophilic EE. In fact, when the drug is encapsulated in the inner core, it is more preserved from lack and degradation.

5.4 Microscopy Characterization

Transmission Electron Microscope (TEM) has been used to confirm the round shape of liposomes section and the presence of the lipidic layer. An example is reported in Fig. 3.

As shown in Fig. 3, the perimeter of liposomes is much evidenced if compared with the white inner core and the external bulk.

Moreover, the total working cost of SuperLip lab-scale plant was estimated. In particular, those calculations were performed on the basis of 1 h of work. First, it was expected that human resources cost more than reagents. Finally, heating, pumping and refrigerating cost was evaluated. All these points of cost were summed together to obtain the total working cost for 1 h of the work performed with SuperLip. Then, the final cost was compared with the selling cost of the vesicles

Fig. 3 Transmission electron microscope on theophylline loaded liposomes

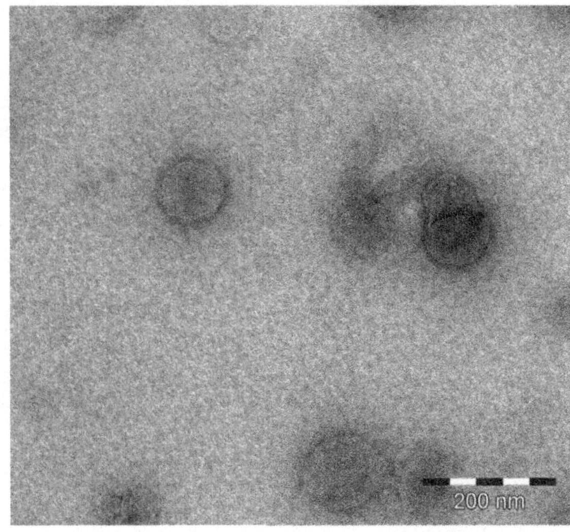

Table 3 Lab scale plant working cost: human resources

Human resources	
Operators	2
One operator cost	35 (€/h)
Total human working cost	70 (€/h)

suspension produced in 1 h, according to an average of prices available online. Human cost is reported in Table 3.

For human resources, it was considered that the presence of two operators is required for health safety. The experiment lasts about an hour, and consequently the total human cost is 70 €/h (see Table 3).

In Table 4 reagent costs used for costs estimation are also reported. The average consumptions of CO_2 has been calculated as an average of all the experiment previously performed with SuperLip since its construction. The average volume consumption of carbon dioxide was also converted into mass consumption, taking into account the average density of CO_2 at the operative pressure and temperature. The quantities of water and ethanol required were fixed, instead. The quantity of lipids required were an average of all the experiments performed in the past, as well as the drug encapsulated into liposomes. Considering the drugs, also the cost is an average of the cost of all the drugs purchased for this work, as reported in Table 4.

For the formation of the water droplets, distilled water was directly self-produced in our laboratories. For this reason, it was considered free for SuperLip unit process. It was evaluated that every single experiments costs 4.16 € in terms of reagent involved. Taking into account the start-up time, as the average experiment time lasts 1 h, the reagents cost is 4.16 €/h.

Table 4 Lab scale plant working cost: reagents

Reagents cost				
Compound	Quantity used		Buying cost	1 exp. cost
CO_2	200 dm^3	374 g	0.60 €/Kg	0.224 €
H_2O	300 mL		/	Self produced
Ethanol	100 mL		26.88 €/L	2.69 €
Lipids	500 mg		2 €/g	1 €
Drug	50 mg		5 €/g	0.25 €
1 experiment reagent cost				4.16 €
Experiment time				1 h
Total experiment cost				4.16 €/h

Table 5 Lab scale plant working cost: heaters (**A**), pumps and refrigeration (**B**) cost

A. Heating cost	
Formation vessel (FV)	2 × 800 W
Saturator (S)	2 × 140 W
Separation step (SS)	1 × 275 W
Total power required	2155 W = 2.155 KW
Power cost in this area	0.3374 €/KWh
Total Power cost	0.7271 €h

B. Pumping and refrigerating costs				
/	H$_2$O pump	Ethanol pump	CO$_2$ pump	Refrigeration bath
Efficiency	0.70	0.70	0.75	0.90
Power (KW)	0.125	0.125	0.370	0.170
1 Exp. Cost (€)	0.060	0.060	0.166	0.064
Total power cost		0.35 €/h		

In Table 5 the heating, pumping and refrigerating cost used for costs estimation are reported. Particularly, for the liquid pumps it was considered a working efficiency of 0.70, while for the carbon dioxide it was estimated a coefficient of 0.75, instead for the refrigeration bath an efficiency of 0.90.

As reported in Table 5, total power cost is greatly less than reagent and human cost. This means that SuperLip does not require great power supply, especially in the idea of scaling up the process. In fact, the total power supply cost is 1.08 €/h, that represents only the 1.44% of the total working cost.

Finally, it is possible to say that SuperLip has total operative cost of 5.24 €/h, without taking into account the human cost.

In Table 6 the net profit calculation is reported.

As shown in Table 6, it was considered that 1 mL of liposomes suspension might be sold at the price of 2.67 €/mL at least, as reported in the online catalogues for similar liposomes drug formulations. Therefore, the suspension produced during

Table 6 Net profit calculation for liposomes sales on a lab scale plant production

Net profit on liposome sales (€/h)	
Human resources	70.00
Reagent cost	4.16
Power heaters	0.73
Power pump and refrigeration	0.35
Total working cost	75.24
Approx. liposomes liquid solution sales revenue	2.67
Gross profit	801.00
Net profit	725.76

1 h with SuperLip at the average water flow rate of 10 mL/min can be sold at the price of 801 €. The gross profit is almost 1064.6% of the total working cost. Considering that the total working cost is 75.24 €/h, SuperLip guarantees a net profit of 725.76 €/h.

6 Conclusions

SuperLip is a powerful technique for the production of liposomes of nanometric mean diameters, exploiting the fast diffusion coefficient of carbon dioxide in supercritical conditions.

Different kind of hydrophilic compounds were successfully incorporated in water droplets with SuperLip process. Moreover, also lipophilic compounds were successfully entrapped inside the double lipidic layer of liposomes. In both cases, the measured encapsulation efficiencies were included between a minimum of 86.2% to a maximum of 99.0%, confirming the higher potential of this technique compared to other conventional and supercritical processes.

In the future, other kind of compounds will be encapsulated to understand if it is possible to choose the optimal operative parameters according to the family that the compounds belong to.

Acknowledgements The authors want to acknowledge Annabella Bruno, Marina Iorio, Luca Vitale and Daniela Apicella for their kind help with SuperLip process during their thesis projects.

References

1. Tazina, E.V., Kostin, K.V., Oborotova, N.A.: Specific features of drug encapsulation in liposomes (a Review). Pharm. Chem. J. **45**(8), 481–490 (2011)
2. Allen, T.M., Cullis, P.R.: Drug delivery systems: entering the mainstream. Science **303**(5665), 1818–1822 (2004)
3. Uhrich, K.E., Cannizzaro, S.M., Langer, R.S., Shakesheff, K.M.: Polymeric systems for controlled drug release. Chem. Rev. **99**(11), 3181–3198 (1999)

4. Farokhzad, O.C., Langer, R.: Impact of nanotechnology on drug delivery. ACS Nano 3(1), 16–20 (2009)
5. Youan, B.B.C.: Impact of nanoscience and nanotechnology on controlled drug delivery. Nanomedicine 3(4), 401–406 (2008)
6. Immordino, M.L., Dosio, F., Cattel, L.: Stealth liposomes: review of the basic science, rationale, and clinical applications, existing and potential. Int. J. Nanomed. 1(3), 297–315 (2006)
7. Xu, X.M., Costa, A., Burgess, D.J.: Protein encapsulation in unilamellar liposomes: high encapsulation efficiency and a novel technique to assess lipid-protein interaction. Pharm. Res. 29(7), 1919–1931 (2012)
8. Li, J.L., Cheng, X.D., Chen, Y., He, W.M., Ni, L., Xiong, P.H., Wei, M.G.: Vitamin E TPGS modified liposomes enhance cellular uptake and targeted delivery of luteolin: an in vivo/in vitro evaluation. Int. J. Pharm. 512(1), 262–272 (2016)
9. Pinto-Alphandary, H., Andremont, A., Couvreur, P.: Targeted delivery of antibiotics using liposomes and nanoparticles: research and applications. Int. J. Antimicrob. Agents 13(3), 155–168 (2000)
10. Heppner, D.G., et al.: Safety, immunogenicity, and efficacy of Plasmodium falciparum repeatless circumsporozoite protein vaccine encapsulated in liposomes. J. Infect. Dis. 174(2), 361–366 (1996)
11. Schwarz, G., Robert, C.H.: Kinetics of pore-mediated release of marker molecules from liposomes or cells. Biophys. Chem. 42(3), 291–296 (1992)
12. Kibria, G., Hatakeyama, H., Sato, Y., Harashima, H.: Anti-tumor effect via passive anti-angiogenesis of PEGylated liposomes encapsulating doxorubicin in drug resistant tumors. Int. J. Pharm. 509(1–2), 178–187 (2016)
13. Gibbs, B.F., Kermasha, S., Alli, I., Mulligan, C.N.: Encapsulation in the food industry: a review. Int. J. Food Sci. Nutr. 50(3), 213–224 (1999)
14. Coutelle, C., Williamson, R.: Liposomes and viruses for gene therapy of cystic fibrosis. J. Aerosol Med.-Deposition Clearance and Eff. Lung 9(1), 79–88 (1996)
15. Dass, C.R., Walker, T.L., DeCruz, E.E., Burton, M.A.: Cationic liposomes and gene therapy for solid tumors. Drug Delivery 4(3), 151–165 (1997)
16. Goyal, K., Huang, L.: Gene therapy using DC-Chol liposomes. J. Liposome Res. 5(1), 49–60 (1995)
17. Lasic, D.D.: Liposomes in gene therapy. Chimica Oggi-Chemistry Today 14(3–4), 13–16 (1996)
18. Miller, A.D.: Cationic liposomes for gene therapy. Angewandte Chemie-Int. Ed. 37(13–14), 1769–1785 (1998)
19. Schuber, F., Kichler, A., Boeckler, C., Frisch, B.: Liposomes: from membrane models to gene therapy. Pure Appl. Chem. 70(1), 89–96 (1998)
20. Iwanaga, K., Ono, S., Narioka, K., Morimoto, K., Kakemi, M., Yamashita, S., Nango, M., Oku, N.: Oral delivery of insulin by using surface coating liposomes—Improvement of stability of insulin in GI tract. Int. J. Pharm. 157(1), 73–80 (1997)
21. Mozafari, M.R.: Liposomes: an overview of manufacturing techniques. Cell. Mol. Biol. Lett. 10(4), 711–719 (2005)
22. Reverchon, E., Adami, R., Campardelli, R., Della Porta, G., De Marco, I., Scognamiglio, M.: Supercritical fluids based techniques to process pharmaceutical products difficult to micronize: Palmitoylethanolamide. J. Supercrit. Fluids 102, 24–31 (2015)
23. Campardelli, R., Baldino, L., Reverchon, E.: Supercritical fluids applications in nanomedicine. J. Supercrit. Fluids 101, 193–214 (2015)
24. Prosapio, V., Reverchon, E., De Marco, I.: Polymers' ultrafine particles for drug delivery systems precipitated by supercritical carbon dioxide plus organic solvent mixtures. Powder Technol. 292, 140–148 (2016)
25. Della Porta, G., Nguyen, B.N.B., Campardelli, R., Reverchon, E., Fisher, J.P.: Synergistic effect of sustained release of growth factors and dynamic culture on osteoblastic

differentiation of mesenchymal stem cells. J. Biomed. Mater. Res., Part A **103**(6), 2161–2171 (2015)

26. Campardelli, R., Reverchon, E.: alpha-Tocopherol nanosuspensions produced using a supercritical assisted process. J. Food Eng. **149**, 131–136 (2015)
27. Maherani, B., Arab-Tehrany, E., Mozafari, M.R., Gaiani, C., Linder, M.: Liposomes: a review of manufacturing techniques and targeting strategies. Curr. Nanosci. **7**(3), 436–452 (2011)
28. Espirito Santo, I., Campardelli, R., Albuquerque, E.C., Vieira de Melo, S., Della Porta, G., Reverchon, E.: Liposomes preparation using a supercritical fluid assisted continuous process. Chem. Eng. J. **249**, 153–159 (2014)
29. Campardelli, R., Trucillo, P., Reverchon, E.: A supercritical fluid-based process for the production of fluorescein-loaded liposomes. Ind. Eng. Chem. Res. **55**(18), 5359–5365 (2016)
30. Fang, J.Y., Lee, W.R., Shen, S.C., Huang, Y.L.: Effect of liposome encapsulation of tea catechins on their accumulation in basal cell carcinomas. J. Dermatol. Sci. **42**(2), 101–109 (2006)
31. Lozada-Castro, J.J., Santos-Delgado, M.J.: Determination of free cholesterol oxide products in food samples by gas chromatography and accelerated solvent extraction: influence of electron-beam irradiation on cholesterol oxide formation. J. Sci. Food Agric. **96**(12), 23–4215 (2016)
32. Campardelli, R., Espirito Santo, I., Albuquerque, E.C., Vieira de Melo, S., Della Porta, G., Reverchon, E.: Efficient encapsulation of proteins in submicro liposomes using a supercritical fluid assisted continuous process. J. Supercrit. Fluids **107**, 163–169 (2016)

Supercritical Antisolvent Process: PVP/Nimesulide Coprecipitates

Iolanda De Marco, Valentina Prosapio and Ernesto Reverchon

Abstract Nimesulide (NIM) is an anti-inflammatory drug, widely used in the treatment of acute pain associated with different diseases. A major limitation in its usage is due to its reduced solubility in water; therefore, large doses are required to reach the therapeutic level, with consequent undesired effects on patient's health. In order to improve NIM dissolution rate, a possible solution is represented by its micronization. Traditional micronization techniques show several drawbacks: lack of control over the particle morphology and particle size distribution, large solvent residues and use of high temperatures. An alternative to conventional techniques is represented by supercritical carbon dioxide ($scCO_2$) based processes. In particular, nanoparticles and microparticles of different kind of materials were successfully obtained by supercritical antisolvent (SAS) precipitation. However, when processed using SAS, nimesulide precipitated in form of large crystals or it is completely extracted by the mixture solvent/antisolvent. A solution to this problem can be the production of drug-polymer composite microspheres, using a water soluble polymer in which the drug is entrapped. In this work, NIM coprecipitation with polyvinylpyrrolidone (PVP) is proposed on pilot scale. The effects of polymer/drug ratio, concentration, pressure and temperature were investigated to identify successful operating conditions for SAS coprecipitation. Microparticles with a mean diameter ranging between 1.6 and 4.1 μm were successfully produced. Drug release analyses revealed that NIM dissolution rate from PVP/NIM microparticles was 2.5 times faster with respect to unprocessed drug. The possible precipitation mechanisms involved in the process were discussed.

Keywords Nanocomposite microparticles · Coprecipitation · Nimesulide · Polyvinylpyrrolidone · Supercritical antisolvent process · Precipitation mechanisms

I. De Marco (✉) · V. Prosapio · E. Reverchon
Department of Industrial Engineering, University of Salerno,
Via Giovanni Paolo II 132, 84084 Fisciano, SA, Italy
e-mail: idemarco@unisa.it

© Springer International Publishing AG 2018
S. Piotto et al. (eds.), *Advances in Bionanomaterials*,
Lecture Notes in Bioengineering, DOI 10.1007/978-3-319-62027-5_4

1 Introduction

Nimesulide (NIM) is a non-steroidal anti-inflammatory drug, largely employed for the treatment of back pain, toothache, postoperative pain and inflammation, head-ache and migraine [1]. A major limitation in its usage is due to its reduced solubility in water (<0.01 mg/mL), which imply the necessary use of large doses to reach the therapeutic level, with consequent undesired effects on patient's health, such as heartburn, nausea, diarrhea, vomiting, peptic ulcer and hepatic damages [2]. In order to improve NIM dissolution rate, and correspondingly reduce its dose, a possible solution is represented by particle size reduction at micrometric diameters thanks to the larger surface area in contact with the medium [3]. Traditional micronization processes, such as spray-drying, emulsification/solvent evaporation, liquid antisolvent precipitation, freeze drying and jet-milling, suffer from some drawbacks: lack of control over the particle morphology and particle size distri-bution, large solvent residues and use of high temperatures [4].

An alternative to conventional techniques is represented by supercritical carbon dioxide (scCO$_2$) based processes, characterized by fast mass transfer, high solvent power, high density, near zero surface tension, low viscosity and high diffusivity, that can be tuned varying pressure and temperature. They have been successfully applied in different fields: micronization [5–7], liposomes' formation [8], extraction [9], impregnation [10], membranes and scaffolds production [11, 12]. In particular, nanoparticles and microparticles of different kind of materials were successfully obtained by supercritical antisolvent (SAS) precipitation [13–15]. However, when processed using SAS, NIM precipitated in form of large crystals or it is completely extracted by the mixture formed by the organic solvent and scCO$_2$, as a result of the partial solubility of this drug in the mixture solvent/antisolvent [16]. In order to overcome this limitation, an effective solution can be the production of drug-polymer composite microspheres, using a water soluble polymer in which the drug is entrapped. The fast solubilization of the polymer should release the drug in nanometric sub-microparticles obtaining an improvement of their bioavailability. Nevertheless, SAS coprecipitation is difficult to obtain, since the two compounds generally tend to precipitate separately. In SAS literature only few papers are devoted to coprecipitates formation and the results showed in these works are characterized by irregular and coalescing particles, with broad particle size distri-butions, low drug entrapment efficiency and, in some cases, questionable demon-stration of the occurred coprecipitation [17–20]. Recently, Reverchon and co-workers obtained successful results in the production of composite micro-spheres Polyvinylpyrrolidone (PVP)/liposoluble vitamins [21], PVP/β-carotene [22] and PVP/corticosteroids [23], taking advantage of PVP ability to retard crystal growth [24]. NIM coprecipitation with PVP was also performed on bench scale [25]: coprecipitated microparticles were obtained, overcoming the above mentioned limitations and, therefore, improving NIM bioavailability.

In this work, NIM coprecipitation with PVP is proposed on pilot scale. The following effects will be tested: polymer/drug ratio, operating pressure, temperature and overall solute concentration. Samples will be characterized using different techniques in order to verify the occurred coprecipitation and the improvement of NIM dissolution rate. An explanation of the involved coprecipitation mechanisms will be proposed.

2 Materials, Methods and Procedures

2.1 Materials

Polyvinylpyrrolidone (PVP, average molecular weight 10 kg/mol), Nimesulide (NIM, purity $\geq$ 98%) and Dimethylsulfoxide (DMSO, purity 99.5%) were supplied by Sigma-Aldrich (Italy). CO_2 (purity 99%) was purchased from Morlando Group s.r.l. (Italy). All materials were used as received. Solubility tests performed at 20 °C showed that: the solubilities of the materials in DMSO are about 250 mg/mL in the case of PVP and 200 mg/mL in the case of NIM.

2.2 SAS Apparatus and Procedure

SAS pilot plant used for the experiments reported in this work is shown in Fig. 1. It consists of two high pressure pumps used to deliver the liquid solution and supercritical CO_2, respectively. A cylindrical vessel of 5.2 dm^3 internal volume (I. V.) (L/D ratio equal to 9.4) is used as the precipitation chamber. The liquid mixture and the supercritical CO_2 are fed to the precipitator through a tube-in-tube injection system (internal tube d = 3 mm; annulus d = 8.5 mm). The generation of small liquid droplets is ensured by the presence of a 0.5 mm nozzle fitted on the tip of the internal tube. The gaseous CO_2 is cooled in a shell and tube heat exchanger, condensed and recirculated to the storage vessel. Pressure is measured by a test gauge manometer and regulated by a micrometric valve located at the exit of the precipitator. Temperature is set by a PID controller connected with electrically thin bands. A stainless steel frit (pore diameter of 0.1 μm) located at the bottom of the precipitator is used to collect the produced powder, allowing the CO_2-solvent-solution to pass through. A second collection vessel located downstream the micrometric valve, whose pressure is regulated by a backpressure valve, is used to recover the liquid solvent. At the exit of the second vessel, CO_2 flow rate and the total quantity of antisolvent delivered are measured by a rotameter and a dry test meter, respectively.

A SAS experiment begins delivering CO_2 to the SAS vessel until the desired pressure is reached; then, pure solvent is sent through the nozzle to the vessel for at least 15 min. When a quasi-steady state composition of solvent and antisolvent is realized inside the SAS vessel, the flow of the solvent is stopped and the liquid solution is delivered through the nozzle, producing the precipitation of the solute.

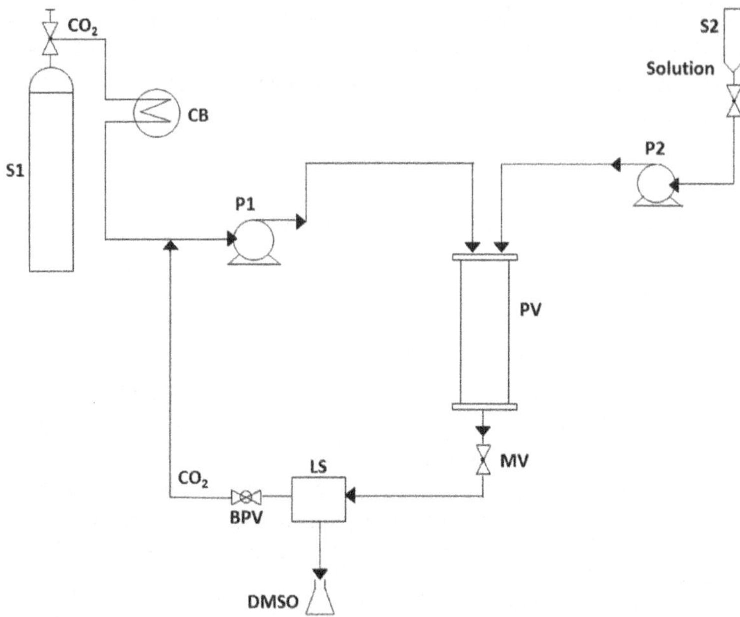

Fig. 1 Schematic representation of SAS apparatus. *S1* CO_2 supply; *S2* liquid solution supply; *CB* cooling bath; *P1*, *P2* pumps; *PV*: precipitation vessel; *MV* micrometering valve; *LS* liquid separator; *BPV* back-pressure valve

At the end of the solution delivery, supercritical CO_2 continues to flow, to wash the precipitator, eliminating the solution formed by the liquid solubilized in the supercritical antisolvent. If the final purge with pure CO_2 is not performed, the organic solvent contained in the fluid phase condenses during the depressurization step and can solubilize or modify the precipitates. At the end of the washing step, CO_2 flow is stopped and the precipitator is depressurized down to atmospheric pressure.

2.3 Analytical Methods

Samples of the precipitated material were observed by a Field Emission Scanning Electron Microscope (FE-SEM, mod. LEO 1525, Carl Zeiss SMT AG, Oberkochen, Germany). Powder was dispersed on a Carbon tab previously stuck to an Aluminum stub (Agar Scientific, United Kingdom); then, was coated with Gold (layer thickness 250 Å) using a sputter coater (mod. 108 A, Agar Scientific, Stansted, United Kingdom).

Particle size distribution (PSD) of the powders was measured from FE-SEM photomicrographs using the Sigma Scan Pro image analysis software (release 5.0, Aspire Software International Ashburn, VA). Approximately 1000 particles, taken at high enlargements and in various locations inside the precipitator, were analyzed in the elaboration of each particle size distribution. Histograms representing the particle size distributions were fitted using Microcal Origin Software (release 8.0, Microcal Software, Inc., Northampton, MA).

Drug dissolution studies were performed using an UV/vis spectrophotometer (model Cary 50, Varian, Palo Alto, CA) at a wavelength of 396 nm. Accurately weighted samples containing an equivalent amount of NIM (1.5 mg) were suspended in 2 mL of phosphate buffered saline solution (PBS) and placed into a dialysis sack; they were then incubated in 300 mL of PBS at pH 7.4, continuously stirred at 200 rpm and 37 °C. Each analysis was carried out in triplicate and the proposed curves are the mean profiles.

Precipitation yield (PY %) of each sample was measured by UV–vis analysis, measuring the absorbance obtained in the release medium at the end of the drug release; i.e., when all NIM was released from the microspheres to the outer water phase. The absorbance was, then, converted into NIM concentration, using a calibration curve. Therefore, PY % was calculated comparing the amount of NIM in the initial solution and the amount of NIM in the SAS processed powder, as shown below:

$$PY\% = \frac{\text{mg NIM processed sample}}{\text{mg NIM initial solution}} \times 100 \qquad (1)$$

DMSO residues were measured using a headspace sampler (model 7694E, Hewlett Packard, USA) coupled to a gas chromatograph equipped with a flame ionization detector (GC-FID, model 6890 GC-SYSTEM, Hewlett Packard, Agilent Technologies Mfg. Gmbh & Co. KG, USA). The solvent was separated using two fused silica capillary columns connected in series by press-fit: the first column (model Carbowax EASYSEP, Stepbios, Italy) connected to the detector, 30 m length, 0.53 mm i.d., 1 μm film thickness and the second (model Cp Sil 5CB CHROMPACK, Stepbios, Italy) connected to the injector; 25 m length, 0.53 mm i. d., 5 μm film thickness. GC conditions were: oven temperature at 160 °C for a total time equal to 8.80 min. The injector was maintained at 250 °C (split mode, ratio 5:1), and helium was used as the carrier gas (2 mL/min). Head space conditions were: equilibration time, 9 min at 170 °C; pressurization time, 0.3 min; loop fill time, 0.4 min. Head space samples were prepared in 20 mL vials filled with 50 mg of drug dissolved in water. Analyses were performed on each batch of processed drug, in triplicates.

3 Experimental Results

SAS experiments were carried out using a CO_2 flow rate equal to 20 kg/h and a solution flow rate equal to 10 mL/min (10 grams of material were processed in each test). DMSO was used as the liquid solvent. The effect of PVP/NIM w/w ratio (R),

Table 1 Summary of SAS experiments performed on PVP/NIM (*MP* microparticles; *CM* coalescing material; *M* morphology; *m.d.* mean diameter; *s.d.* standard deviation; *PY %* precipitation yield)

#	P, MPa	R, w/w	T, °C	C, mg/mL	M	m.d.μm	s.d., μm	PY %
1.	9	0:1	40	30	–	–	–	–
2.	9	1:0	40	30	MP	4.80	2.32	–
3.	9	2:1	40	30	CM	–	–	65.8
4.	9	3:1	40	30	MP	2.27	1.33	98.4
5.	9	5:1	40	30	MP	3.92	1.81	99.8
6.	9	10:1	40	30	MP	4.03	2.16	99.5
7.	9	5:1	40	20	MP	3.98	2.20	99.1
8.	9	5:1	40	35	MP	3.96	2.12	99.8
9.	9	5:1	35	30	MP	1.62	0.81	99.2
10.	9	5:1	45	30	MP	4.12	2.68	98.9
11.	15	5:1	40	30	MP	3.12	1.62	96.6

Fig. 2 FE-SEM image of PVP precipitated from DMSO at 9 MPa, 40 °C, 30 mg/mL

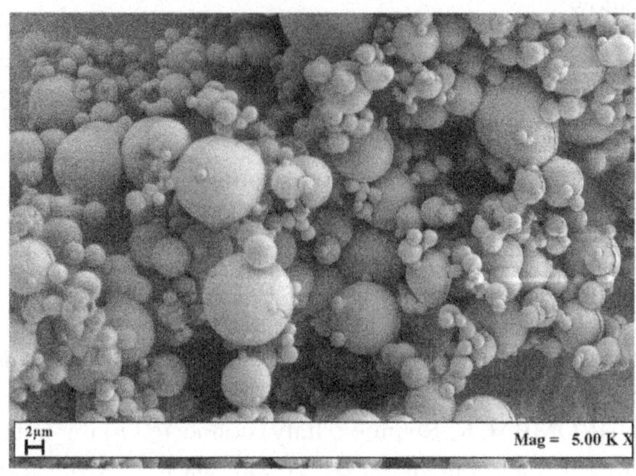

total concentration (C), operating pressure (P) and temperature (T) was investigated. A list of the most relevant experiments was reported in Table 1, with the indication of the operating conditions employed, the obtained morphology and, in the case of particles, their mean diameter (m.d.) and standard deviation (s.d.).

The first experiment was performed processing NIM alone (which has a characteristic yellow color) at 9 MPa, 40 °C and 30 mg/mL (run #1 in Table 1); however, no powder was recovered at the end of the experiment since NIM was probably extracted from the mixture CO_2/DMSO.

Then, PVP alone (characterized by a white color) was processed at the same conditions (#2 in Table 1); in this case, well separated microparticles with a mean size equal to 4.80 μm were obtained, as shown in Fig. 2, where a FE-SEM image is reported. Therefore, this polymer represents a good candidate as carrier for SAS coprecipitates formation.

In order to verify if PVP is able to avoid NIM extraction during SAS and to allow a successful coprecipitation, several experiments were performed on the pilot plant on the system PVP/NIM at different process conditions.

3.1 Effect of Polymer/Drug Ratio

The first effect taken into account was the polymer/drug ratio, which was varied from 2:1 to 10:1, at 9 MPa, 40 °C and 30 mg/mL. When PVP/NIM 2:1 (#3 in Table 1) was processed, precipitates were characterized by crystals and coalescing material, suggesting that probably the two compounds precipitated individually. Increasing PVP/NIM ratios at 3:1 (#4), 5:1 (#5) and 10:1 (#6), spherical microparticles were obtained, as shown in the exemplificative images reported in Fig. 3a, b. Moreover, the powder in these last experiments had a uniform yellow color, suggesting the presence of both NIM and PVP.

Comparing the PSDs of PVP/NIM particles, it was observed that increasing the polymer/drug ratio the particle mean size increased and the particle size distribution enlarged, as it is possible to observe from data reported in Table 1.

3.2 Effect of Overall Solute Concentration

The effect of the concentration of solutes in DMSO was investigated at 9 MPa, 40 °C and PVP/NIM 5:1, varying it from 20 to 35 mg/mL. Using a concentration of

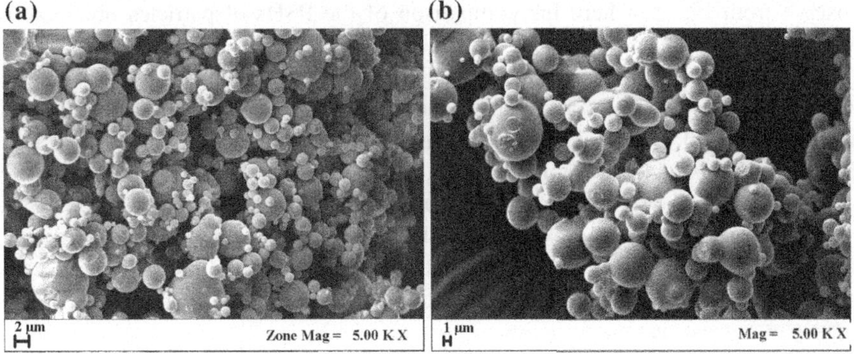

Fig. 3 FE-SEM images of PVP/NIM particles precipitated from DMSO at 9 MPa and 40 °C at different polymer/drug ratios: **a** 5:1; **b** 10:1

                                                I. De Marco et al.

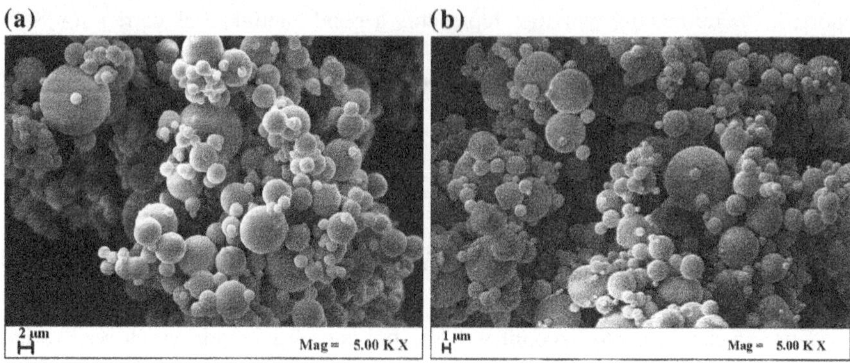

Fig. 4 FE-SEM images of PVP/NIM 5/1 particles precipitated at 9 MPa and 40 °C at different total concentrations in DMSO: **a** 20 mg/mL, **b** 35 mg/mL

20 mg/mL (#7 in Table 1), microparticles, with a mean diameter of about 4 µm, were produced, as showed in Fig. 4a. Increasing the concentration at 35 mg/mL (#8), microparticles were still produced, as it is possible to observe in Fig. 4b.

The comparison of the PSD of the microparticles obtained at 20, 30 and 35 mg/mL, showed that, varying the concentration, no significant change in the particle mean diameter occurred (see Table 1).

3.3 Effect of the Operating Temperature and Pressure

The influence of the temperature on particles morphology and size was studied varying it from 35 to 45 °C. When the temperature was fixed at 35 °C (run #9), microspheres were obtained; increasing the temperature at 45 °C (#10), microparticles were still produced, but with larger mean diameter, as it is possible to observe from Fig. 5, where the comparison of the PSDs of particles obtained at different temperature is reported.

In order to obtain a complete screening of the effect of the SAS operating parameters on PVP/NIM microspheres, the influence of the operating pressure was then investigated. Therefore, an experiment was carried out at 15 MPa (#11 in Table 1). At this condition, microspheres, with a mean diameter of 3.12 µm, were still obtained as shown in Fig. 6, where a FE-SEM image is reported.

Comparing the data reported in Table 1, it is possible to observe that increasing the operating pressure from 9 to 15 MPa, the obtained particles underwent only to a small reduction of the mean size, whilst the morphology did not change.

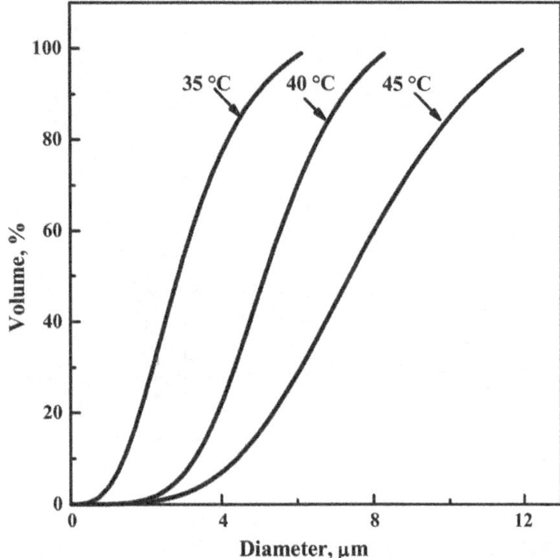

Fig. 5 Volumetric cumulative PSDs of PVP/NIM 5:1 microparticles precipitated from DMSO at 9 MPa, 30 mg/mL and different temperatures

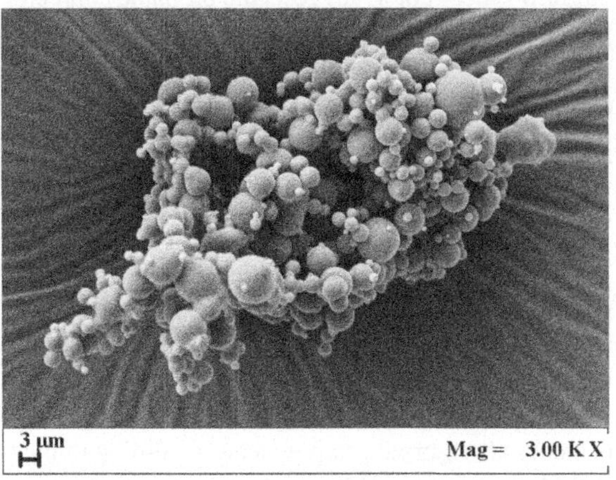

Fig. 6 FE-SEM image of PVP/NIM 5:1 particles precipitated from DMSO at 15 MPa, 40 °C and 30 mg/mL

3.4 *Characterization of Precipitates*

UV-vis spectroscopy was used to measure the precipitation yield (PY %) and the drug dissolution profiles.

PY was found to be equal to 65.8% when coalescing material and crystal were produced. PY, instead, was found to be between 98.4 and 99.8% for all the samples formed by microspheres PVP/NIM, as shown in Table 1; therefore, in presence of PVP, NIM is not extracted by the mixture solvent/antisolvent and it is all present in the powder recovered at the end of the experiment. This evidence, along with the

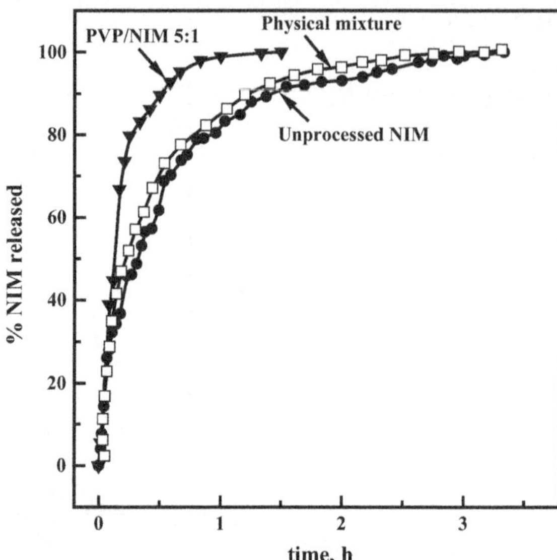

Fig. 7 Dissolution profiles of NIM in PBS at 37 °C and pH 7.4

uniform yellow color, suggests the probable coprecipitation when microparticles are produced.

In order to have a further evidence, dissolution tests in PBS at 37 °C were performed for unprocessed NIM, the physical mixture PVP/NIM 5/1 and the SAS processed PVP/NIM 5/1 obtained at 9 MPa, 40 °C and 30 mg/mL (run #5 in Table 1). The dissolution rate of the examined samples was monitored for 3.5 h by plotting the percentage of dissolved NIM as a function of time. As shown in Fig. 7, unprocessed NIM and the physical mixture PVP/NIM 5/1 reached the 100% dissolution in about 3.5 h; SAS processed PVP/NIM 5:1 microparticles, instead, showed a reduction of the dissolution time, achieving the complete release in about 1.5 h. Therefore, the dissolution rate of the drug, when entrapped in the microspheres, was largely improved: 2.5 times faster than the unprocessed Nimesulide.

A headspace sampler, coupled to a GC, was used to quantify the solvent residue content in SAS produced powders. According to the Food and Drug Administration (FDA) guidelines, DMSO belongs to class 3 solvents; therefore, its maximum acceptable concentration in the final product is 5000 ppm. The analysis revealed that in all the samples PVP/NIM the solvent residue was around 500 ppm that is largely below the limit fixed by the FDA.

4   Discussion

A possible interpretation of the results discussed in this work can be proposed on the ground of SAS precipitation mechanisms. It has been frequently observed that, during SAS processing, when the operating point is located near above the mixture critical point (MCP) of the binary system solvent/antisolvent, jet break-up and drying mechanism takes place, with consequent formation of microparticles. When the solution is formed by two compounds, since all the content of the droplet concurs to form the dried particle, coprecipitation is successful. This hypothesis has been confirmed by PY and dissolution tests, which showed the presence of NIM in the processed powder and an improvement of NIM dissolution rate of 2.5 times, compared to the unprocessed drug. This last observation could be explained hypothesizing that NIM is formed by uniformly distributed nanoparticles inside each PVP microparticle. Moreover, it was observed that it is possible to modulate particle mean size, according to the employed operating conditions.

The study on the effect of the polymer/drug ratio showed that, starting from PVP/NIM 3/1, microspheres are produced and that increasing the ratio the particle mean diameter increases as well. In correspondence of 2/1, instead, the presence of coalescing material was observed, suggesting that the precipitation behavior of PVP is still not predominant, which is also confirmed by the lower PY. Varying the temperature, it was observed that increasing this parameter the particle mean size increased and the PSD widened. This result is due to a broadening of the miscibility gap of the system $DMSO/CO_2$ at higher temperatures [26]; therefore, the operating point become closer to the MCP, with consequent precipitation of larger particles. Varying the pressure, it was observed that at 15 MPa microparticles are still produced, but their mean size decreased. The obtainment of this morphology is uncommon, since at this pressure, the system should be in the single phase mixing, with production of nanoparticles. However, this phenomenon has already been observed for PVP by De Marco et al. [27]: also at high pressure it continues to produce microparticles.

5 Conclusions

SAS process demonstrated to be effective in polymer/drug coprecipitation, also on pilot scale, when selected operating conditions are employed. The use of PVP as carrier revealed to be a key strength since it allowed not only to avoid NIM extraction, but also to obtain well separated spherical microparticles with improved drug dissolution rate.

References

1. Pouchain, E.C., Costa, F.W.G., Bezerra, T.P., Soares, E.C.S.: Comparative efficacy of nimesulide and ketoprofen on inflammatory events in third molar surgery: a split-mouth, prospective, randomized, double-blind study. International Journal of Oral and Maxillofacial Surgery **44**(7), 876–884 (2015). doi:10.1016/j.ijom.2014.10.026
2. Dashora, K., Saraf, S., Saraf, S.: Effect of processing variables and in-vitro study of microparticulate system of nimesulide. Revista Brasileira de Ciências Farmacêuticas **43**, 555–562 (2007)

3. Saffari, M., Ebrahimi, A., Langrish, T.: A novel formulation for solubility and content uniformity enhancement of poorly water-soluble drugs using highly-porous mannitol. Eur J Pharm Sci **83**, 52–61 (2016). doi:10.1016/j.ejps.2015.12.016

4. Wang, W., Liu, G., Wu, J., Jiang, Y.: Co-precipitation of 10-hydroxycamptothecin and poly (l-lactic acid) by supercritical CO2 anti-solvent process using dichloromethane/ethanol co-solvent. The Journal of Supercritical Fluids **74**, 137–144 (2013). doi:10.1016/j.supflu.2012.11.022

5. Couto, R., Alvarez, V., Temelli, F.: Encapsulation of Vitamin B2 in solid lipid nanoparticles using supercritical CO2. J. Supercrit. Fluids **120**, Part 2, 432–442 (2017). doi:10.1016/j.supflu.2016.05.036

6. Prosapio, V., Reverchon, E., De Marco, I.: Antisolvent micronization of BSA using supercritical mixtures carbon dioxide + organic solvent. J. Supercrit. Fluids **94**, 189–197 (2014)

7. Reverchon, E., Adami, R., De Marco, I., Laudani, C.G., Spada, A.: Pigment Red 60 micronization using supercritical fluids based techniques. J. Supercrit. Fluids **35**(1), 76–82 (2005)

8. Campardelli, R., Trucillo, P., Reverchon, E.: A supercritical fluid-based process for the production of fluorescein-loaded liposomes. Ind. Eng. Chem. Res. **55**(18), 5359–5365 (2016). doi:10.1021/acs.iecr.5b04885

9. Hossain, M.S., Norulaini, N.A.N., Naim, A.Y.A., Zulkhairi, A.R.M., Bennama, M.M., Omar, A.K.M.: Utilization of the supercritical carbon dioxide extraction technology for the production of deoiled palm kernel cake. J. CO$_2$ Util. **16**, 121–129 (2016). doi:10.1016/j.jcou.2016.06.010

10. Smirnova, I., Mamic, J., Arlt, W.: Adsorption of drugs on silica aerogels. Langmuir **19**(20), 8521–8525 (2003)

11. Baldino, L., Concilio, S., Cardea, S., De Marco, I., Reverchon, E.: Complete glutaraldehyde elimination during chitosan hydrogel drying by SC-CO$_2$ processing. J. Supercrit. Fluids **103**, 70–76 (2015). doi:10.1016/j.supflu.2015.04.020

12. Baldino, L., Concilio, S., Cardea, S., Reverchon, E.: Interpenetration of natural polymer aerogels by supercritical drying. Polymers **8**(4), 106 (2016). doi:10.3390/polym8040106

13. Badens, E., Majerik, V., Horváth, G., Szokonya, L., Bosc, N., Teillaud, E., Charbit, G.: Comparison of solid dispersions produced by supercritical antisolvent and spray-freezing technologies. Int. J. Pharm. **377**(1), 25–34 (2009)

14. De Marco, I., Reverchon, E.: Supercritical carbon dioxide + ethanol mixtures for the antisolvent micronization of hydrosoluble materials. Chem. Eng. J. **187**, 401–409 (2012). doi:10.1016/j.cej.2012.01.135

15. Rueda, M., Sanz-Moral, L.M., Segovia, J.J., Martín, Á.: Enhancement of hydrogen release kinetics from ethane 1,2 diamineborane (EDAB) by micronization using Supercritical Antisolvent (SAS) precipitation. Chem. Eng. J. **306**, 164–173 (2016). doi:10.1016/j.cej.2016.07.052

16. Moneghini, M., Perissutti, B., Vecchione, F., Kikic, I., Alessi, P., Cortesi, A., Princivalle, F.: Supercritical antisolvent precipitation of nimesulide: preliminary experiments. Curr. Drug Deliv. **4**(3), 241–248 (2007). doi:10.2174/156720107781023901

17. Montes, A., Gordillo, M.D., Pereyra, C., Martínez de la Ossa, E.J.: Co-precipitation of amoxicillin and ethyl cellulose microparticles by supercritical antisolvent process. J. Supercrit. Fluids **60**, 75–80 (2011). doi:10.1016/j.supflu.2011.05.002

18. Kurniawansyah, F., Mammucari, R., Foster, N.R.: Inhalable curcumin formulations by supercritical technology. Powder Technol. **284**, 289–298 (2015). doi:10.1016/j.powtec.2015.04.083

19. Jin, H.Y., Xia, F., Zhao, Y.P.: Preparation of hydroxypropyl methyl cellulose phthalate nanoparticles with mixed solvent using supercritical antisolvent process and its application in co-precipitation of insulin. Adv. Pow. Tech. **23**(2), 157–163 (2012). doi:10.1016/j.apt.2011.01.007

20. Franceschi, E., De Cezaro, A., Ferreira, S.R.S., Kunita, M.H., Muniz, E.C., Rubira, A.F., Oliveira, J.V.: Co-precipitation of beta-carotene and bio-polymer using supercritical carbon dioxide as antisolvent. Open Chem. Eng. J. **5**(1), 11–20 (2011)
21. Prosapio, V., Reverchon, E., De Marco, I.: Incorporation of liposoluble vitamins within PVP microparticles using supercritical antisolvent precipitation. J. CO_2 Util. **19**, 230–237 (2017)
22. Prosapio, V., Reverchon, E., De Marco, I.: Coprecipitation of Polyvinylpyrrolidone/ β-Carotene by supercritical antisolvent processing. Ind. Eng. Chem. Res. **54**(46), 11568–11575 (2015). doi:10.1021/acs.iecr.5b03504
23. Prosapio, V., De Marco, I., Reverchon, E.: PVP/corticosteroid microspheres produced by supercritical antisolvent coprecipitation. Chem. Eng. J. **292**, 264–275 (2016). doi:10.1016/j.cej.2016.02.041
24. Ledet, G.A., Graves, R.A., Glotser, E.Y., Mandal, T.K., Bostanian, L.A.: Preparation and in vitro evaluation of hydrophilic fenretinide nanoparticles. Int. J. Pharm. **479**(2), 329–337 (2015). doi:10.1016/j.ijpharm.2014.12.052
25. Prosapio, V., Reverchon, E., De Marco, I.: Formation of PVP/nimesulide microspheres by supercritical antisolvent coprecipitation. J. Supercrit. Fluids **118**, 19–26 (2016)
26. Andreatta, A.E., Florusse, L.J., Bottini, S.B., Peters, C.J.: Phase equilibria of dimethyl sulfoxide (DMSO) + carbon dioxide, and DMSO + carbon dioxide + water mixtures. J. Supercrit. Fluids **42**(1), 60–68 (2007). doi:10.1016/j.supflu.2006.12.015
27. De Marco, I., Rossmann, M., Prosapio, V., Reverchon, E., Braeuer, A.: Control of particle size, at micrometric and nanometric range, using supercritical antisolvent precipitation from solvent mixtures: application to PVP. Chem. Eng. J. **273**, 344–352 (2015)

PLA-Based Nanobiocomposites with Modulated Biodegradation Rate

Valentina Iozzino, Felice De Santis, Valentina Volpe
and Roberto Pantani

Abstract The disposal of polymeric waste is increasingly becoming an issue of international concern. The use of biodegradable polymers is a possible strategy to face most of the problems related to the disposal of the durable (non-biodegradable) polymers. Among biodegradable polymers, polylactic acid (PLA), obtained from renewable sources, is a very attractive one, due to its relatively good processability, biocompatibility, interesting physical properties. Hydrolysis is the major depolymerization mechanism and the rate-controlling step of PLA biodegradation in compost. The propensity to degradation in the presence of water significantly limits specific industrial applications such as automotive, biomedical, electronic and electrical appliances, agriculture. Therefore the control of biodegradation rate is somewhat even more important than the characteristic of biodegradability itself. In this scenario, it is critical to find additives able to modulate the biodegradation rate of biodegradable polymers, in relationship to the expected lifetime. Since the kinetics of hydrolysis strongly depend on the pH of the hydrolyzing medium, in this work some fillers able to control the pH of water when it diffuses inside the polymer were added to PLA. In particular, fumaric acid, a bio- and eco- friendly additive, and magnesium hydroxide, a common antiacid, were used. These fillers were added to the material using a melt-compounding technique, suitable for industrial application. The results obtained are encouraging toward the possibility of effectively controlling the degradation rate.

Keywords Polylactic acid · PLA · Nanocomposite · Hydrolytic degradation · Magnesium hydroxide · Fumaric acid

V. Iozzino (✉) · F. De Santis · V. Volpe · R. Pantani
Department of Industrial Engineering, University of Salerno,
Via Giovanni Paolo II 132, 84084 Fisciano, SA, Italy
e-mail: viozzino@unisa.it

© Springer International Publishing AG 2018
S. Piotto et al. (eds.), *Advances in Bionanomaterials*,
Lecture Notes in Bioengineering, DOI 10.1007/978-3-319-62027-5_5

1 Introduction

Over the years, polymers have found an increasingly important use in different sectors. The problem of the use of traditional polymers, beyond the necessity of using oil as a raw material, is their difficult to dispose after use. A very good alternative is the use of biodegradable polymers: their macromolecules, under exposition to atmospheric agents, break down into CO_2, H_2O and inorganic non-toxic material in a lapse of time of a few months. Biodegradation process consists in two different steps: hydrolysis, in which water molecules break down the polymer chains into simpler molecules, and effective biodegradation, in which microorganisms feed on the hydrolysis products of transforming them into compounds that return to be a part of the natural cycle (CO_2, CH_4, biomass).

PLA is one of the most interesting biodegradable polymers, it is an aliphatic polyester derived from 100% renewable sources (synthetic product from starch derived from maize) [1]. Several studies investigated peculiar aspects of this resin, going from crystallization kinetics [2] to processing like injection molding [3, 4] and melt compounding [5–7]. Some investigations have focused on the hydrolytic degradation of aliphatic polyesters, analyzing the factors which can influence such phenomenon [8–10]: it is generally accepted that, among other key factors like temperature, the hydrolysis of PLA in solution depends on the media pH [11, 12]. The propensity of PLA to degradation in the presence of water significantly limits specific industrial applications, particularly for durable products with long-term performance.

In this work, the influence of acid and basic fillers on the hydrolysis rate is explored.

2 Materials

Polylactic acid (PLA) (M_n = 181 kg/mol, M_w = 320 kg/mol) was purchased from Nature Works LLC in pellet form (PLA 4060D). This commercial PLA grade contains about 15% of D-lactide. Because of this relatively high percentage of D-lactide this neat resin is a fully amorphous grade, generally extruded or coextruded to obtain PLA film. Fumaric acid, $C_4H_4O_4$, and magnesium hydroxide, $Mg(OH)_2$, were purchased from Sigma Aldrich.

3 Experimental

3.1 Sample Preparation

The PLA pellets and all the fillers were dried for 24 h under vacuum at the temperature of 60 °C. The materials were melt compounded in a Minilab Haake

Thermoscientific twin-screw mini-extruder, with a homogeneous temperature of 170 °C and at 100 rpm, with a cycle time of 5 min. The materials for subsequent testing were obtained: PLA extruded; PLA +3% wt fumaric acid; PLA +3% wt magnesium hydroxide.

From all the extruded materials several amorphous films were obtained by compression molding (T = 170 °C) with the thickness in the range 200–300 μm.

The consequence of hydrolysis process on these polymer films was monitored following the time evolution of selected sample properties:

- pH change, by the means of a pH-meter, in the solution in which the sample hydrolyzes;
- weight loss of the sample;
- crystallinity degree of the sample.

3.2 Hydrolysis Tests

Hydrolysis process can be schematized in two step. In the first step the water gets into the polymer bulk, breaking the bonds in the amorphous phase and, as a result, the macromolecular chain is divided into fragments shorter than the initial one, soluble in water. In general, for a semi-crystalline PLA, this phenomenon involves initially the amorphous phase with a decrease of molecular weight, without an evident change of the mechanical properties given by the crystalline phase.

In the second step, the molecular weight reduction is followed by the drop of mechanical properties, while the water is still breaking the macromolecular chains: when the products of this degradation become soluble in water, i.e., enough short, a decrease in the weight is observed.

In this work, hydrolysis tests were conducted following the degradation of ten film samples in bi-distilled water (pH = 5.5) at 58 °C. This temperature was carefully chosen weighed because this is the temperature adopted for biodegradation tests according to ASTM and ISO standards [13, 14]. Several films (having a thickness of about 250 μm, weighing about 50 mg and dimensions 1.5×1.5 cm^2) of each of the samples were placed in a glass vessel containing the water. The volume of water [ml] was fixed in the ratio 800:1 with the mass [g] of the dry samples, according to a previously defined protocol [15]. At selected days of hydrolysis, the vessels containing the samples were emptied by sucking the liquid with a syringe, dried under vacuum at 60 °C for 2.5 h and then weighed. One of the films of each sample, after measuring its weight, was placed in a desiccator for further analysis. All the other vessels were re-filled with water and put again at 58 ° C to continue the hydrolysis test. The hydrolysis of the samples was followed for 21 days.

3.3 *Calorimetric Tests (DSC)*

A differential scanning calorimetry analysis was carried out on all the samples. The mass of the samples for these tests was about 6 mg. The tests were carried out by means of a Mettler Toledo DSC 822 under a nitrogen atmosphere. The samples were heated from −10 to 200 °C at 10 °C/min, held for 2 min at 200 °C, cooled from 200 to 10 °C and then heated again from −10 to 200 °C. The thermal analysis was conducted to evaluate the glass transition temperature (T_g) and the crystalline degree (X_c) [16].

4 Results

4.1 *Hydrolysis Tests: PH Evolution*

The evolution of pH during hydrolysis test in water is shown in Fig. 1.

During the first week, a large deviation in the measured pH of the water is observed, for all the samples, before the hydrolysis starts.

This behavior is consistent with the degradation process in which the water molecules are present in thermodynamic equilibrium as

$$2H_2O \rightleftharpoons H_3O^+ + OH^- \tag{1}$$

At first, the water penetrates inside the polymeric matrix: the ions hydronium H_3O^+ attack the polymeric macromolecules and are, therefore, taken from the solution.

This scenario implies a decrease of ions hydronium concentration in water: the solution is thus richer in OH^- ions and as a result there is an increase of the pH.

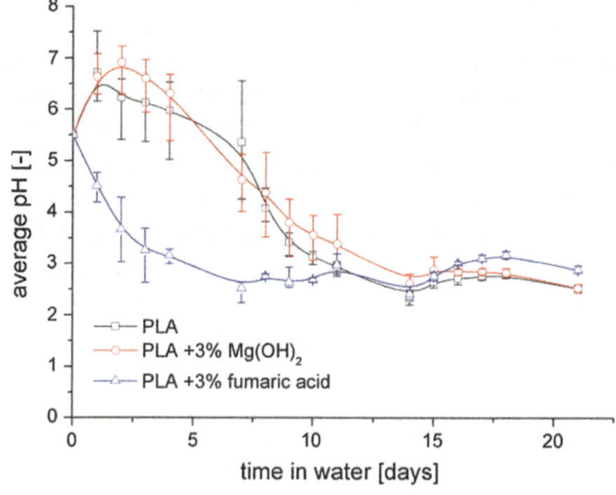

Fig. 1 Evolution of average pH of the water during hydrolysis at 58 °C

With the advancement of the hydrolysis process, the counter-diffusion of the reaction products (carboxylic acids, alcohol and H^+ ions) that pass into solution is observed: this involves the passing of ions in solution H^+. When the hydrolysis is completed the release of H^+ ions stops, therefore the measured pH is found to be almost unchanged compared to that of the starting solution.

For a short time, the subtraction of H^+ ions from solution prevails, while for long times the hydrolysis production of H^+ and in water the related release overcomes. As it can be observed from the data collected experimentally in Fig. 1, for the samples of PLA the pH value grows for the first days and then dropping until it reached a practically constant value. In particular, the samples with the addition of 3% of fumaric acid present an evident drop of pH: the fumaric acid in the samples is the reason of this behavior with a rapid release of H^+ to the water, reaching a constant value of 3 just after the first week.

4.2 Hydrolysis Tests: Weight and Appearance Evolution

In the hydrolysis process, with water diffusion in the polymer and subsequent depolymerization, the soluble fragments of the macromolecular chain go into water solution with gradual reduction of sample weight.

The evolution of the average residual weight is reported in Fig. 2. The samples of PLA and PLA with 3% of $Mg(OH)_2$ present a minimum variation of the weight during the first ten days, while after this period a relevant and rapid reduction is evident, reaching 50% of starting value after 21 days.

As for pH evolution, the samples of PLA with 3% of fumaric acid present a different behavior: the weight shows a sharper variation after the first week so that at the end of the test, after 21 days, the samples have a weight that is the 25% of the starting one.

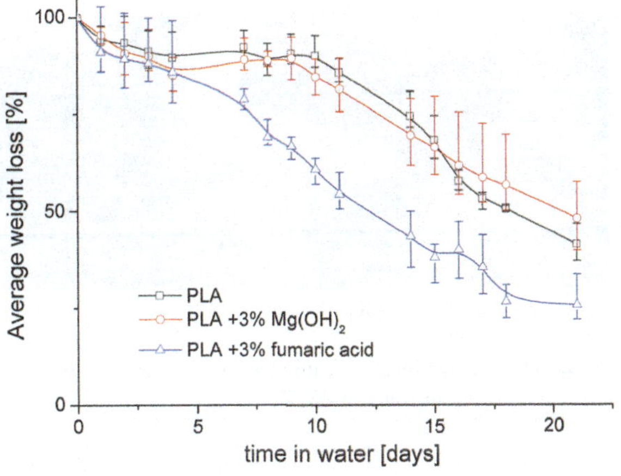

Fig. 2 Evolution of average residual weight during hydrolysis at 58 °C

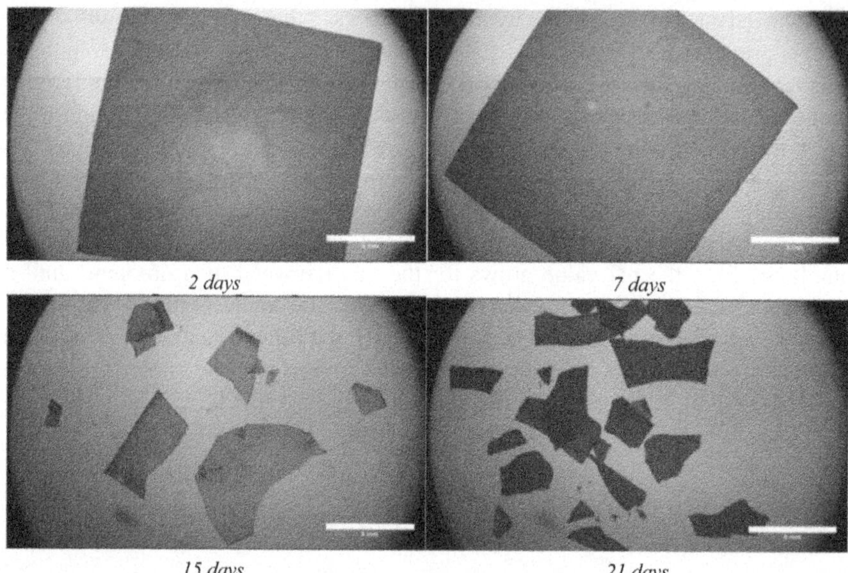

Fig. 3 Microscopy images, with scale bar 5 mm, of the PLA samples during hydrolysis in water at 58 °C

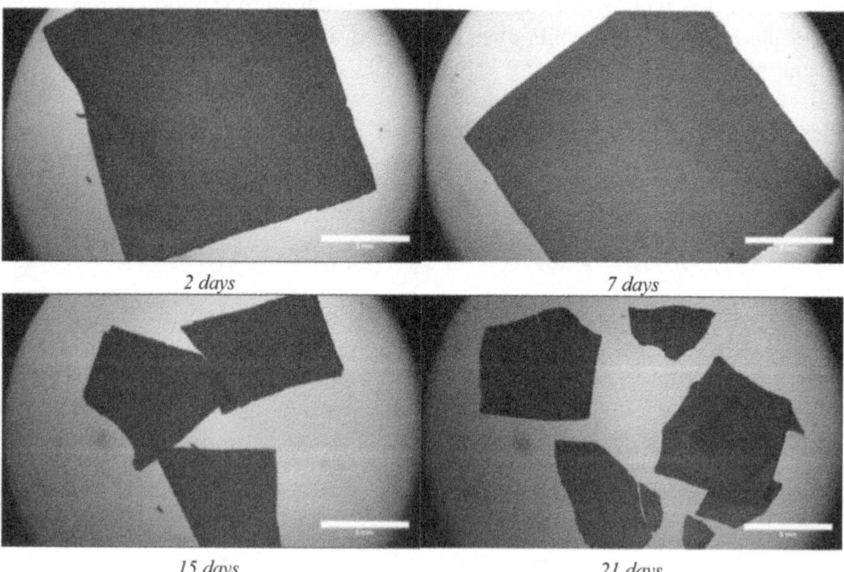

Fig. 4 Microscopy images, with scale bar 5 mm, of the PLA with 3% of Mg(OH)$_2$ samples during hydrolysis in water at 58 °C

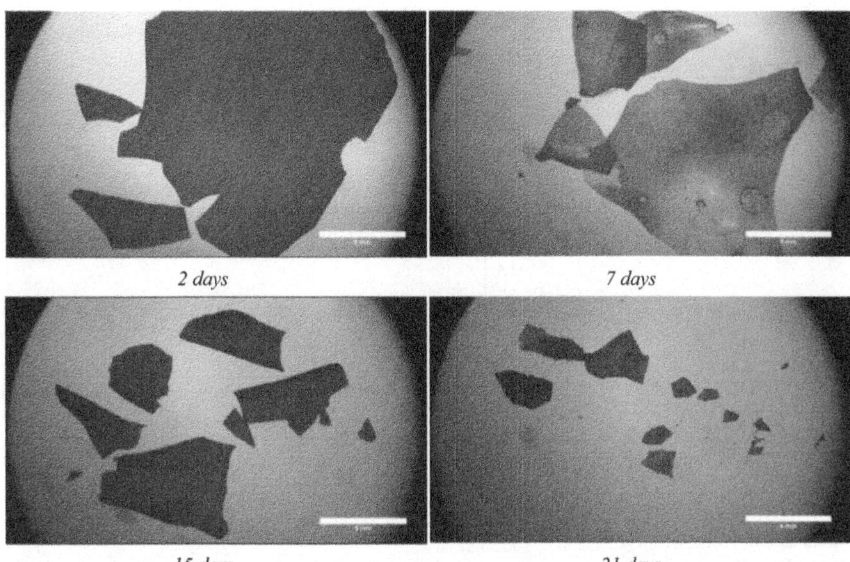

Fig. 5 Microscopy images, with scale bar 5 mm, of the PLA with 3% of fumaric acid samples during hydrolysis in water at 58 °C

The weight evolution is confirmed also by the appearance of the samples, shown in Figs. 3, 4, and 5. Also observing the images it is clear that the addition of fumaric acid to PLA 4060D induces a more rapid degradation and fragmentation of the material. This phenomenon can be explained by the fact that the hydrolysis is an auto-catalytic process that leads to the production of acidic products (-COOH) in the terminal part of the chain. Due to the slow diffusion of the terminal of the chains, the acids fragments may accumulate within the chain and cause a faster degradation of the macromolecules.

On the contrary, the addition of magnesium hydroxide seems to have a weak retarding effect on the process, although these changes, if compared to the pure material behavior, are limited.

4.3 Hydrolysis Tests: Calorimetric Characterization of the Samples

The thermal analysis was conducted for the evaluation of the glass transition temperature (T_g) and the crystalline degree (X_c).The samples (melt compounded PLA, with fumaric acid and with magnesium hydroxide) hydrolyzed to 15, 50, and 70% were analyzed.

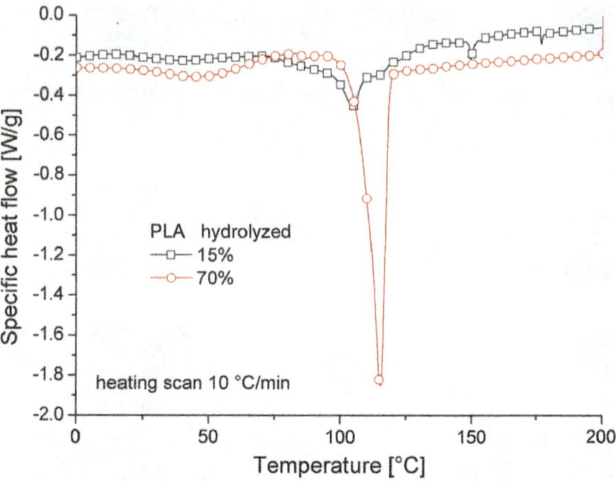

Fig. 6 Thermograms, during the first heating scan at 10 °C/min, of hydrolyzed PLA film samples

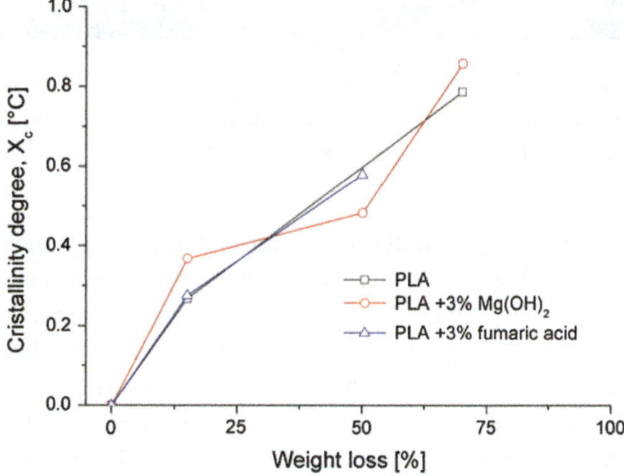

Fig. 7 Crystallinity degree, evaluated from first heating scan at 10 °C/min, of hydrolyzed film samples

As expected, all the film samples are completely amorphous (Fig. 7), before the hydrolysis test, with similar glass transition temperature of about 54–56 °C (Fig. 8) that is also the same of the neat unprocessed 4060D PLA resin [17, 18].

During the hydrolysis test, as for the other characterizations, the calorimetric results show an interesting behavior, as shown as an example in Fig. 6 for the neat resin. The first heating reveals a melting peak related to the melting of the crystalline phase, absent in the not hydrolyzed samples. It is confirmed by calorimetric characterization in all the hydrolyzed samples, and crystallinity degree is higher for longer hydrolysis time. This behavior can be related to the fragmentation of macromolecules, and as a consequence, the crystallinity degree is almost linearly linked to the weight loss, as shown in Fig. 7.

Obviously similar crystallinity degree, about 0.5, corresponding to the same weight loss, 50%, is reached by the sample with fumaric acid (15 days) in less time than the neat sample or with magnesium hydroxide (22 days).

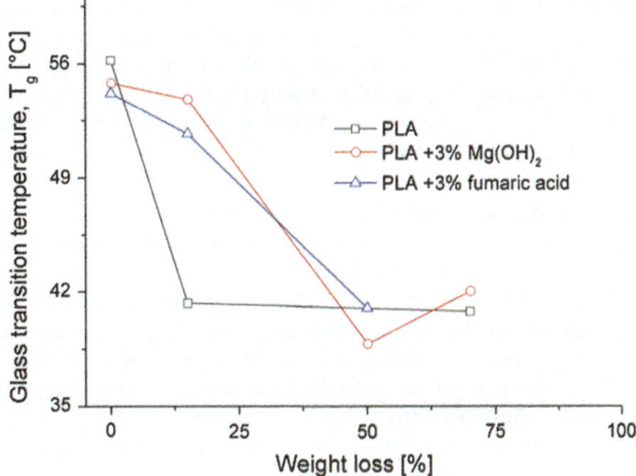

Fig. 8 Glass transition temperature, evaluated from cooling scan at 10 °C/min, of hydrolyzed film samples

On the other hand, glass transition temperature depression is evident in all film samples on increasing of weight loss, as shown in Fig. 8. Once again this phenomenon can be related to increased mobility with de-polymerization of macromolecules in hydrolysis tests.

5 Conclusions

Since the kinetics of hydrolysis strongly depend on the pH of the hydrolyzing medium, in this work some fillers able to control the pH of water when it diffuses inside the polymer were added to PLA. In particular, fumaric acid, a bio- and eco-friendly additive, and magnesium hydroxide, a common antacid, were used. These fillers were added to the material using the melt-compounding technique, suitable for industrial application. The results obtained are encouraging toward the possibility of accelerating the degradation rate with fumaric acid. On the other side, the protective effect of magnesium hydroxide is less evident, if compared to the neat PLA hydrolysis test. The degradation, during hydrolysis, affects the pH of the water in which the test is conducted and the sample weight. Therefore the fragmentation of macromolecular chain has a strong influence on crystallization of the samples so that it changes this resin, that usually is amorphous, increasing crystallinity degree with hydrolyzation.

References

1. Garlotta, D.: A literature review of poly (lactic acid). J. Polym. Environ. **9**(2), 63–84 (2001)
2. De Santis, F., Pantani, R., Titomanlio, G.: Nucleation and crystallization kinetics of poly (lactic acid). Thermochim. Acta **522**(1), 34–128 (2011)

3. Pantani, R., De Santis, F., Sorrentino, A., De Maio, F., Titomanlio, G.: Crystallization kinetics of virgin and processed poly (lactic acid). Polym. Degrad. Stab. **95**(7), 59–1148 (2010)
4. De Santis, F., Volpe, V., Pantani, R.: Effect of molding conditions on crystallization kinetics and mechanical properties of poly (lactic acid). Polym. Eng. Sci. **57**(3), 306–311 (2017)
5. De Santis, F., Pantani, R.: Melt compounding of poly (Lactic Acid) and talc: assessment of material behavior during processing and resulting crystallization. J. Polym. Res. **22**(12), 1–9 (2015)
6. Concilio, S., Iannelli, P., Sessa, L., Olivieri, R., Porta, A., De Santis, F., et al.: Biodegradable antimicrobial films based on poly (lactic acid) matrices and active azo compounds. J. Appl. Polym. Sci. **132**(33) (2015)
7. Sessa, L., Concilio, S., Iannelli, P., De Santis, F., Porta, A., Piotto, S. (ed.): Antimicrobial azobenzene compounds and their potential use in biomaterials. International advances in applied physics and materials science congress & exhibition (APMAS'15): Proceedings of the 5th International Advances in Applied Physics and Materials Science Congress & Exhibition, AIP Publishing (2016)
8. Pantani, R., Sorrentino, A.: Influence of crystallinity on the biodegradation rate of injection-moulded poly (lactic acid) samples in controlled composting conditions. Polym. Degrad. Stab. **98**(5), 96–1089 (2013)
9. Lostocco, M.R., Huang, S.J.: The hydrolysis of poly (lactic acid)/poly (hexamethylene succinate) blends. Polym. Degrad. Stab. **61**(2), 30–225 (1998)
10. Henton, D.E., Gruber, P., Lunt, J., Randall, J.: Polylactic acid technology. Nat. Fibers, Biopolymers, Biocomposites **16**, 77–527 (2005)
11. De Jong, S., Arias, E.R., Rijkers, D., Van, Nostrum C., Kettenes-Van den Bosch, J., Hennink, W.: New insights into the hydrolytic degradation of poly (lactic acid): participation of the alcohol terminus. Polymer **42**(7), 802–2795 (2001)
12. Tsuji, H., Nakahara, K.: Poly (L-lactide). IX. Hydrolysis in acid media. J. Appl. Polym. Sci. **86**(1), 94–186 (2002)
13. Pantani, R., De Santis, F. (eds.): Physical changes of poly (lactic acid) induced by water sorption. Polymer processing with resulting morphology and properties: feet in the present and eyes at the future. Proceedings of the GT70 International Conference, AIP Publishing (2015)
14. Pantani, R., De Santis, F., Auriemma, F., De Rosa, C., Di Girolamo, R.: Effects of water sorption on poly (lactic acid). Polymer **99**, 9–130 (2016)
15. Gorrasi, G., Pantani, R.: Effect of PLA grades and morphologies on hydrolytic degradation at composting temperature: assessment of structural modification and kinetic parameters. Polym. Degrad. Stab. **98**(5), 14–1006 (2013)
16. De Santis, F., Gorrasi, G., Pantani, R.: A spectroscopic approach to assess transport properties of water vapor in PLA. Polym. Testing **44**, 15–22 (2015)
17. Duan, Z., Thomas, N.L.: Water vapour permeability of poly (lactic acid): crystallinity and the tortuous path model. J. Appl. Phys. **115**(6), 064903 (2014)
18. Davis, E.M., Minelli, M., Baschetti, M.G., Sarti, G.C., Elabd, Y.A.: Nonequilibrium sorption of water in polylactide. Macromolecules **45**(18), 94–7486 (2012)

Part II
Modelling of Bionanomaterials

Modelling Approach to Enzymatic pH Oscillators in Giant Lipid Vesicles

Ylenia Miele, Tamás Bánsági Jr., Annette F. Taylor and Federico Rossi

Abstract The urease-catalyzed hydrolysis of urea can display feedback driven by base production (NH_3) resulting in a switch from acidic to basic pH under non-buffered conditions. Thus, this enzymatic reaction is a good candidate for investigation of chemical oscillations or bistability. In order to determine the best conditions for oscillations, a two-variable model was initially derived in which acid and urea were supplied at rates k_H and k_S from an external medium to an enzyme-containing compartment. Oscillations were theoretically observed providing the necessary condition that $k_H > k_S$ was met. To apply this model, we devised an experimental system able to ensure the fast transport of acid compared to that of urea. In particular, by means of the *droplet transfer method*, we encapsulated the enzyme, together with a proper pH probe, in 1-palmitoyl-2-oleoyl-sn-glycero-3-phosphatidylcholine (POPC) based liposomes, where differential diffusion of H^+ and urea is ensured by the different permeability (P_m) of the membrane to the two species. Here we present an improved theoretical model that accounts for the products transport and for the probe hydrolysis, to obtain a better guidance for the experiments.

Keywords Enzymatic oscillators · Urea-urease reaction · Lipid vesicles · pH oscillators

Y. Miele · F. Rossi (✉)
Department of Chemistry and Biology, University of Salerno, Via Giovanni Paolo II 132, 84084 Fisciano, SA, Italy
e-mail: frossi@unisa.it

T. Bánsági Jr. · A.F. Taylor
Department of Chemical and Biological Engineering, University of Sheffield, Mappin Street, S1 3JD Sheffield, UK

© Springer International Publishing AG 2018

S. Piotto et al. (eds.), *Advances in Bionanomaterials*,
Lecture Notes in Bioengineering, DOI 10.1007/978-3-319-62027-5_6

"

1 Introduction

Oscillations can spontaneously arise in the concentration of some chemical reaction intermediates, provided that the system is sufficiently far from the state of the thermodynamic equilibrium [1]. The first examples of oscillating reactions can be traced back to the beginning of the 20th century, however it is only in the mid 1960's, after the discovery and characterization of the Belousov-Zhabotinsky (BZ) reaction that chemical oscillations begun to attract attention [2]: over the years the BZ system has been performed in a great variety of reaction environments such as gels [3], micelles [4–9], water in oil reverse microemulsions [10–13], lipid bilayers [14–18], organic and inorganic beads [19, 20] always looking for new spatio-temporal behaviours. Moreover, the number of chemical oscillators grew so much that it is now possible to list almost 200 variants ranging from the bromate family to the Mn-based to the most recent family of pH oscillators [2].

Oscillations in pH occur naturally in plants, during glycolysis in yeast cells and are also used by C. elegans for muscle contraction. In these systems the pH oscillations have small amplitude and are driven by other autocatalytic processes, rather than a direct feedback through acid/base interaction; on the contrary, in *synthetic* pH oscillators, the hydrogen ion plays the most important role in the reaction kinetics and the variation in pH can be as large as 6 pH-units [2]. So far several pH oscillators have been developed, however all of them require open (flow) reactors and harsh inorganic redox chemistry; in contrast to this, there is a need for increasing the range of closed autocatalytic reactions, in mild or physiological conditions, for biocompatible applications and pH oscillators driven by enzyme-catalyzed reactions seem to be a valid and promising alternative [21].

Among the several enzymatic reactions known to depend on the pH, one candidate, the urea-urease system has recently been revised by our group [21–23]. The hydrolysis of urea by the enzyme urease results in the production of a weak base, ammonia, and carbon dioxide, which hydrolyses to give bicarbonate. This reaction occurs in numerous cellular systems, for example it is used by bacteria *Helicobacter Pylori* in order to raise the local pH to protect himself from the harsh acidic environment of the stomach [24]. The urea-urease reaction follows Michaelis-Menten kinetics and has a bell-shaped rate-pH curve with maximum at pH 7. In non-buffered conditions, the rate-pH curve can be exploited to obtain feedback-driven behaviour, e.g. by delivering an acid or a base to the solution a reaction acceleration or inhibition can be obtained. The reaction has also the distinct advantages of high solubility and stability of substrate and enzyme in water making it suitable for experiments in vitro. Few nonlinear dynamical behaviors were obtained in different experimental conditions (bistability, front propagation, etc.) [21, 25], however oscillations were never observed.

In a recent work [23], we showed that for obtaining oscillations a fast transport of the acid with respect to that of urea must be ensured in the reaction medium; we then proposed to use 1-palmitoyl-2-oleoyl-sn-glycero-3-phosphatidylcholine (POPC) liposomes to encapsulate the enzyme and profit of the different membrane permeability (P_m) of H^+ ($\sim 10^{-3}$ cm/s) and urea ($\sim 10^{-6}$ cm/s) [26, 27].

The encapsulation of the enzyme into liposomes could be attained by means of the droplet transfer method, introduced by the Weitz group [28], recently optimised by Luisi and collaborators [29] and successfully employed to confine the Belousov-Zhabotinsky chemical oscillator in a network of vesicles [30]. This innovative method first takes advantage of the facile compartmentalization of water-soluble solutes (enzyme in this case) in water-in-oil (w/o) droplets, and then convert the solute-filled w/o droplets into vesicles that can be dispersed in an acidic solution of urea. The pH changes in time can be monitored through the use of a pH probe, pyranine (8-hydroxy-1,3,6-pyrenetrisulfonate) whose emission intensity (at 510 nm) is strongly dependent upon the pH of the solution, over the range 6–10. Both the probe and the enzyme do not leak out of the vesicles once entrapped therein, while urea and proton can easily diffuse through the membrane at different rates (depending on the concentration gradient between the inside and the outside of the membrane).

The diffusion of urea (k_S) and proton (k_H) through the lipid bilayer has been investigated experimentally and the transport of acid was found to be faster than urea, thus confirming that the necessary condition for oscillation $k_H > k_S$ is met. Experimental investigations also showed that liposomes are indeed *open* compartments where both the reactants and the products can permeate the bilayer membranes by a passive diffusion mechanism; thus, the original two-variable model (see Eqs. (1) and (2) in the Results section) which did not account for the leakage of the products [23], had to be modified. In this work we present an improved model which accounts for the ammonia leakage and for the hydrolysis of the pH fluorescent probe used in the experiments; simulations show new phase diagrams where oscillations are still present but in a different part of the plane. Also other dynamical behaviours, such as low and high pH steady states, autocatalysis, clock regimes, etc. were predicted by the model for more realistic experimental conditions.

2   Results: Modelling

The two-variable model for describing the temporal dynamics of the urea-urease reaction in a confined environment, where a solution of urea (S) and acid (H^+) could be delivered into the system, had the form:

$$\frac{d[S]}{dt} = k_S \left([S]_0 - [S]\right) - R \tag{1}$$

$$\frac{d[H^+]}{dt} = \left(k_H\left([H^+]_0 - \frac{K_w}{[H^+]_0} - [H^+] + \frac{K_w}{[H^+]}\right) - 2R\right)\left(1 + \frac{K_w}{[H^+]^2}\right)^{-1} \tag{2}$$

where

$$R = \frac{k_E[\text{E}][\text{S}]}{(K_M + [\text{S}])\left(1 + \frac{K_{es2}}{[\text{H}^+]} + \frac{[\text{H}^+]}{K_{es1}}\right)} \tag{3}$$

$k_E = 3.7 \times 10^{-6}$ ml Mu^{-1}s^{-1}, $K_m = 3 \times 10^{-3}$ M, $K_{ES1} = 5 \times 10^{-6}$ M, $K_{ES2} = 2 \times 10^{-9}$ M are urease specific quantities and $K_w = 10^{-14}$ M^2 is the ion product of water.

The rate constants for the hydrolysis equilibria and for the enzymatic rate constants were taken from the literature, with values slightly adapted to match our experimental conditions [21]. The kinetic constants of the passive transport are expressed as $k_i = 6P_i/d$ where P_i denotes the permeability coefficients ($P_S = 2.45 \times 10^{-6}$ cm/s and $P_H = 1.82 \times 10^{-3}$ cm/s for bilayers made of lipids having a 16 carbon long hydrocarbon chain [27]) and $d = 10$ μm is chosen as the diameter of the vesicle.

The model was scaled and simulations were carried out with XPPAUT [31] using CVODE as integration method; the diagrams reported in Fig. 1 were obtained by using the routine AUTO.

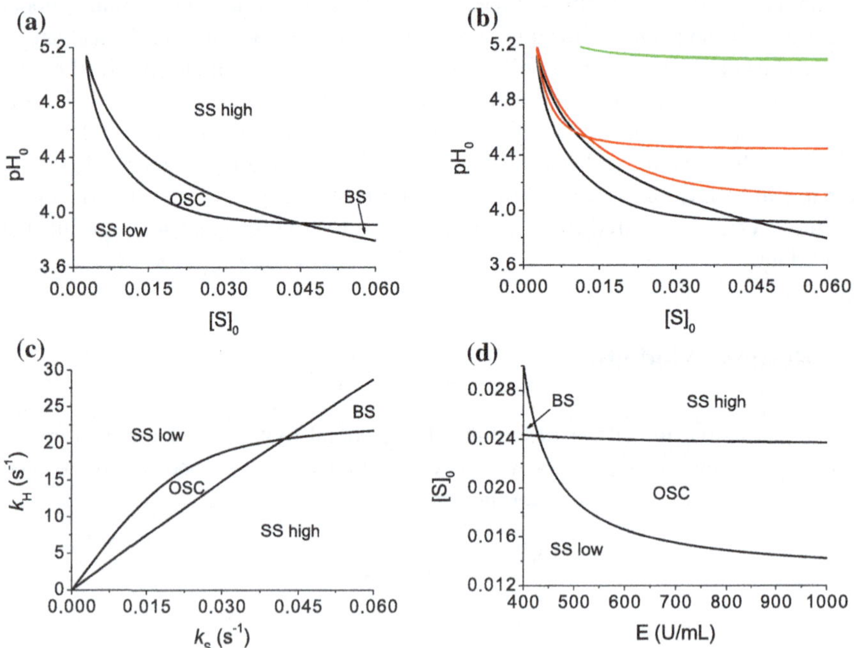

Fig. 1 **a** Phase diagram for E = 1300 U/mL, $k_S = 0.0147$ s^{-1} $k_H = 10.9$ s^{-1}. **b** Phase diagrams for E = 1300 (–), 130 (–), 13 (–), U/ml. **c** Phase diagram for E = 1300 U/mL, pH$_0$ = 4.1, [S]$_0$ = 0.02 M. **d** $k_S = 0.0147$ s^{-1}, $k_H = 10.9$ s^{-1}, pH$_0$ = 4.2

The cross-shaped phase diagrams reveal the presence of four possible states: two stable states (SS low and SS high) and two unstable (a bistability region and an oscillatory domain). The phase diagrams in Fig. 1a and b were obtained in the pH_0 versus $[S]_0$ parameter space and Fig. 1b shows the effect of the enzyme concentration on the shape of the diagram: decreasing the enzyme content the region of oscillations diminishes while the region of bistability increases; for very low concentrations (e.g. E = 13 U/mL) the regions of bistability and oscillations become both very thin.

Figure 1c depicts the effect of the transport constants and, indirectly, the effect of the vesicle size (see the relationship between the permeability and the diameter) on the dynamical behaviour: for some droplet sizes the oscillations were suppressed, but a large region where oscillations exist is present.

Finally in Fig. 1d the phase diagram was obtained in the $[S]_0$ versus E parameter space once the value of pH_0 is fixed.

Despite the good results obtained with the two-variable model, we decided to add few more processes taking place in the vesicles to improve the confidence for the ability of the simulations to guide experiments.

As in the case of the two-variable model, here we take into account the main reactions occurring in the urea-urease system (without including the carbon dioxide equilibria) but we do not consider the steady state approximation for ammonia/ammonium equilibrium, we add the hydrolysis of the pH probe pyranine (pyrOH) and, most important, we consider that reaction products can leak out from the vesicles. The new model has, therefore, the form:

$$H^+_{out} \overset{k_H}{\rightleftharpoons} H^+ \tag{4}$$

$$CO(NH_2)_{2\,out} \overset{k_S}{\rightleftharpoons} CO(NH_2)_2 \tag{5}$$

$$CO(NH_2)_2 + H_2O \overset{R}{\rightarrow} 2NH_3 + CO_2 \tag{6}$$

$$NH_3 + H^+ \overset{k_2}{\underset{k_{2r}}{\rightleftharpoons}} NH_4^+ \tag{7}$$

$$H^+ + OH^- \overset{k_5}{\underset{k_{5r}}{\rightleftharpoons}} H_2O \tag{8}$$

$$pyrOH \overset{k_f}{\underset{k_r}{\rightleftharpoons}} H^+ + pyrO^- \tag{9}$$

$$NH_3 \overset{k_N}{\rightleftharpoons} NH_{3\,out} \tag{10}$$

A cell-like compartment that reproduces our experimental conditions is schematically drawn in Fig. 2: the enzyme and the probe are encapsulated inside the vesicles, while the substrate and the acid are delivered as external solutions.

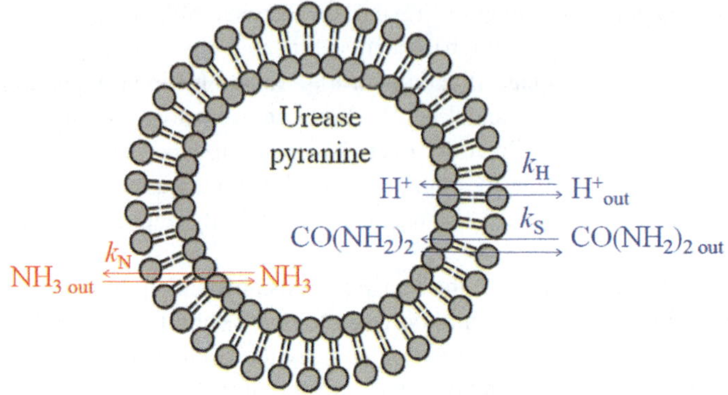

Fig. 2 Lipid vesicle containing the urea-urease reaction and the fluorescence probe: k_H = exchange rate of the acid, k_S = exchange rate of the urea, k_N = exchange rate of the product ammonia with the external solution of concentrations $[H^+]_o$, $[S]_o$ and $[NH_3]_o$

The exchange of matter between the urease-loaded compartment and its surrounding is modelled by the flow terms (4), (5) and (10). In Fig. 2 the substances added from the outside are shown in blue, while ammonia that is produced inside is shown in red.

The system of ODEs derived from reaction kinetics reads:

$$\frac{d[S]}{dt} = k_S([S]_o - [S]) - R \tag{11}$$

$$\frac{d[NH_3]}{dt} = 2R + k_{2r}[NH_4^+] - k_2[NH_3][H^+] + k_N([NH_3]_o - [NH_3]) \tag{12}$$

$$\frac{d[H^+]}{dt} = k_{2r}[NH_4^+] - k_2[NH_3][H^+] + k_f[pyrOH] - k_r[pyrO^-][H^+]$$
$$+ k_H([H^+]_o - [H^+]) + k_{5r} - k_5[OH^-][H^+] \tag{13}$$

$$\frac{d[OH^-]}{dt} = k_{5r} - k_5[OH^-][H^+] \tag{14}$$

$$\frac{d[pyrOH]}{dt} = -k_f[pyrOH] + k_r[pyrO^-][H^+] \tag{15}$$

$$\frac{d[pyrO^-]}{dt} = k_f[pyrOH] - k_r[pyrO^-][H^+] \tag{16}$$

$$\frac{d[NH_4^+]}{dt} = k_2[NH_3][H^+] - k_{2r}[NH_4^+] \tag{17}$$

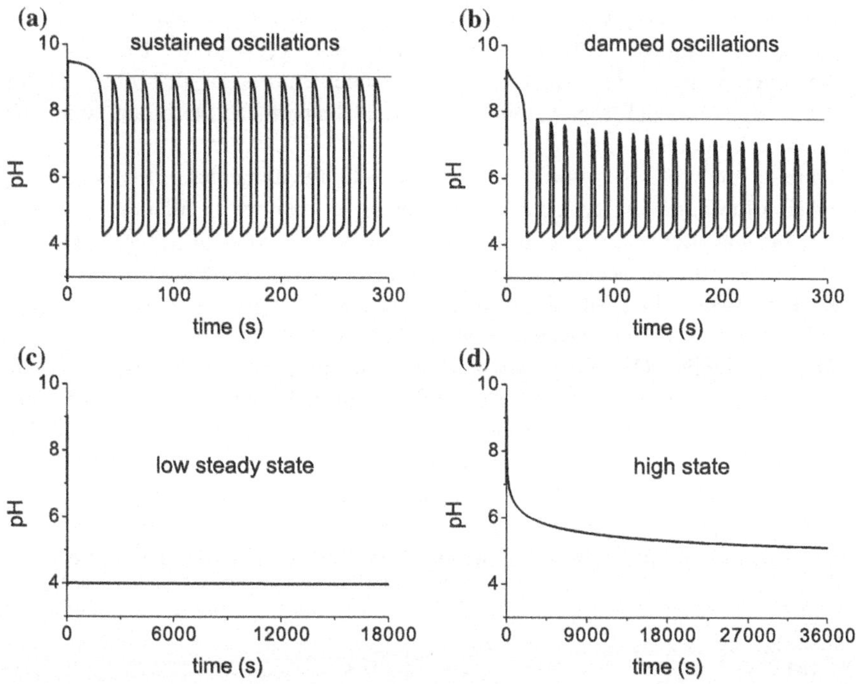

Fig. 3 pH-time plots for E = 1300 U/mL, $k_S = 0.0147$ s^{-1}, $k_H = 10.9$ s^{-1}, $k_N = 1.08$ s^{-1}, $[S]_0 = 0.015$ M, $[NH_3]_0 = 0$ M. **a** with the two-variable model and pH$_0$ = 4.2; **b** with the full model at pH$_0$ = 4.2; **c** with the full model and at pH$_0$ = 3.9; **d** with the full model at pH$_0$ = 4.8. The red line is intended to emphasise the difference between damped and sustained oscillations

The numerical simulations of the Eqs. (11)–(17) were performed using XPPAUT with the same integration method applied for the two-variable model.

Figure 3 shows the most common dynamical behaviours obtained from the integration of the two models; in contrast to the two-variable model which admits sustained oscillations (Fig. 3a), the full system shows damped oscillations (Fig. 3b), i.e. oscillations whose amplitude decreases over time until a steady state is attained. Figure 3c shows a typical low pH steady state, that can be easily found for several experimental parameters. High pH (basic) steady state are, on the contrary, difficult to obtain; this is probably due to the leakage of the NH$_3$ out of the vesicle. In fact, high pH transient states (*high state*) can be obtained for a limited amount of time, but the final attractor of the model is generally a low pH state (see Fig. 3d where the pH is still decreasing over a timescale of 10 h).

The phase diagrams reported in Fig. 4 were obtained by choosing a time interval of 10 h, the initial concentrations of the species inside the vesicle were in the range of those used for the experiments (e.g. pyrOH = 50 μM, S = 0 and $[H^+] = 10^{-7}$ M). The shape of the diagrams is similar to the two-variable model, however some differences can be pointed out: the region of bistability present in Fig. 1a disappears in Fig. 4a and b, being replaced by the oscillatory domain; In Fig. 4c, where the ammonia permeability has been set to a lower value ($k_N = 1.08 \times 10^{-3}$ s^{-1}), the phase diagram E versus [S]$_0$ presents the same features of that obtained for the two-variable model (Fig. 1d), thus the higher the permeability of ammonia, the more the two models differ.

Panels a and b of Fig. 4 show the effect of k_N on the dynamics of the system: by increasing the ammonia permeability the region of damped oscillations becomes wider, but the number of oscillations decreases as reported in Table 1. Finally, Fig. 4d shows the k_S versus k_N parameter space once E, S$_0$ and pH$_0$ have been fixed. Here, when the permeability of ammonia is high, just a low steady state is possible and high pH values cannot be reached.

Despite kinetics (11)–(17) give a more realistic picture with respect to the two-variable model, some processes occurring in the experimental system were still neglected: for example the carbon dioxide equilibria and the leakage of the ammonium ion were not considered. As for the ammonium, we considered that the lipid membrane is only slightly permeable to cations different from protons [27], however, to better understand some of the dynamical behaviours found in simulations (damped oscillations and high states), we further included the permeability

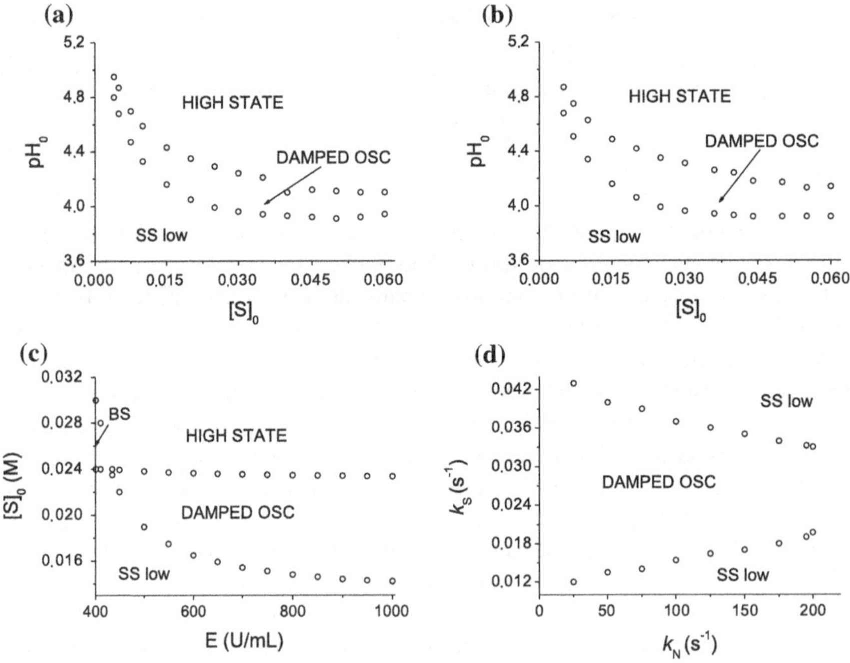

Fig. 4 Phase diagram for E = 1300 U/mL, $k_S = 0.0147$ s^{-1}, $k_H = 10.9$ s^{-1} **a** $k_N = 1.08$ s^{-1}. **b** $k_N = 10.8$ s^{-1}. **c** pH$_0$ = 4.2, $k_S = 0.0147$ s^{-1}, $k_H = 10.9$ s^{-1}, $k_N = 1.08 \times 10^{-3}$ s^{-1}. **d** E = 600 U/mL, [S]$_0$ = 0.02 M, pH$_0$ = 4.2, $k_H = 10.9$ s^{-1}

Table 1 Total number of oscillations for the following conditions: $k_S = 0.0147\,\text{s}^{-1}$, $k_H = 10.9\,\text{s}^{-1}$, $[S]_0 = 0.02$ M, $pH_0 = 4.2$, E = 510 U/mL

Ammonia Permeability (k_N in s^{-1})	Total number of oscillations
0.8	34
1	27
2	13
5	5
7	3

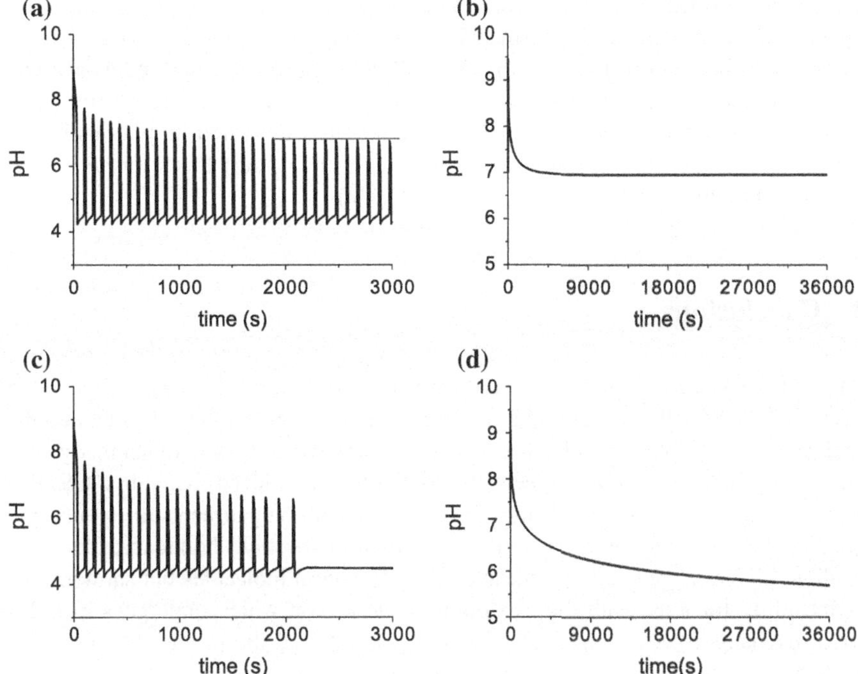

Fig. 5 Simulated time series for $[S]_0 = 0.024$ M, $k_S = 0.0147\,\text{s}^{-1}$, $k_H = 10.9\,\text{s}^{-1}$, $k_N = 1.08\,\text{s}^{-1}$, **a** $k_{NH} = 6\times10^{-4}\,\text{s}^{-1}$, $pH_0 = 4.18$, E = 500 U/mL. **b** $k_{NH} = 6\times10^{-4}\,\text{s}^{-1}$, $pH_0 = 4.2$, E = 450 U/mL. **c** $k_{NH} = 6\times10^{-8}\,\text{s}^{-1}$, $pH_0 = 4.18$, E = 500 U/mL. **d** $k_{NH} = 6\times10^{-8}\,\text{s}^{-1}$, $pH_0 = 4.2$, E = 450 U/mL

of ammonium ion (k_{NH}). A relevant ammonium permeability can be physically possible when the membrane has structural defects and percolation takes place.

Thus, in this case, the following equilibrium must be added to the reactions (4)–(10).

$$\text{NH}_4^+ \overset{k_{NH}}{\rightleftharpoons} \text{NH}_{4\,\text{out}}^+ \qquad\qquad (18)$$

and the ODE for the ammonium ion becomes:

$$\frac{d[NH_4^+]}{dt} = k_2[NH_3][H^+] - k_{2r}[NH_4^+] + k_{NH}([NH_4^+]_o - [NH_4^+]) \qquad (19)$$

If the ammonium permeability (k_{NH}) has a value close or few orders of magnitude lower than the permeability of ammonia (k_N), the system shows the features typical of the two-variable model (possibility of sustained oscillations and high steady states, see Fig. 5a and b), while a low ammonium permeability (e.g. 6×10^{-8} s^{-1}) gives the same results seen in the full model that didn't include the ammonium permeability (Fig. 5c and d show the same trends of Fig. 3b and d).

By properly balancing the permeabilities of ammonium and ammonia, the system behaves like an open reactor in which all the species can enter and exit; a low or zero ammonium permeability makes the system more similar to a semi-batch reactor in which the chemical species are continuously delivered in, but there is no outlet (in this case, the ammonia leaks out easily in contrast to its ionized counterpart, ammonium, that remains entrapped inside).

3　Conclusions

The pH dependence of the urea-urease reaction can be exploited to give base-catalyzed feedback. A simple model with two variables, the substrate and the acid as source of negative feedback, can capture the temporal dynamics of the urea-urease in CSTR but not in cell-like compartments. Indeed, while in a CSTR it is possible to set the flow rate of the reactants, in microcompartments like giant lipid vesicles the flow rate depends on the properties of the lipid boundary: the phospholipid bilayer is generally permeable to small neutral molecules but shows a low permeability for ions, with the exception of the proton which undergoes to a different exchange mechanism [32]. The transport rates of the products, completely neglected in the two-variable model, become relevant in the new model and, actually, they drive the whole dynamical behaviour. In fact, depending on the ratio between the ammonium and the ammonia permeability the model (11)–(17), (19) behaves in two different ways: at higher values of ammonium permeability sustained oscillations and both stable steady states can be possible, at lower values damped oscillations and only a low steady state are observed.

So far in our experiments, we verified the ability of urea, acid and ammonia to cross the lipid bilayer of giant vesicles, furthermore we successfully encapsulated the enzyme and the probe. The next step is to add a solution including both urea and acid to the GVs and see if clock reactions, oscillations or bistability can be found. For this purpose, the simulations presented in this paper could help us in optimising the experimental conditions.

Acknowledgements Y.M. and F.R. were supported by the grants ORSA158121 and ORSA167988 funded by the University of Salerno (FARB ex 60%). The authors acknowledge the support through the COST Action CM1304 (Emergence and Evolution of Complex Chemical Systems).

References

1. Nicolis, G., Prigogine, I.: Self-organization in nonequilibrium systems. Wiley, New York (1977)
2. Orban, M., Kurin-Csorgei, K., Epstein, I.R.: pH-Regulated chemical oscillators. Acc. Chem. Res. **48**(3), 593–601 (2015)
3. Takeoka, Y., Watanabe, M., Yoshida, R.: Self-sustaining peristaltic motion on the surface of a porous gel. J. Am. Chem. Soc. **125**(44) (2003) 13320–13321 pH Oscillators in GVs 11
4. Paul, A.: Observations of the effect of anionic, cationic, neutral, and zwitterionic surfactants on the Belousov-Zhabotinsky reaction. J. Phys.Chem. B **109**(19), 9639–9644 (2005)
5. Rossi, F., Varsalona, R., Liveri, M.L.T.: New features in the dynamics of a ferroincatalyzed Belousov-Zhabotinsky reaction induced by a zwitterionic surfactant. Chem. Phys. Lett. **463** (4–6), 378–382 (2008)
6. Jahan, R.A., Suzuki, K., Mahara, H., Nishimura, S., Iwatsubo, T., Kaminaga, A., Yamamoto, Y., Yamaguchi, T.: Perturbation mechanism and phase transition of AOT aggregates in the Fe (II)[batho(SO3)2]3—catalyzed aqueous Belousov-Zhabotinsky reaction. Chem. Phys. Lett. **485**(4–6), 304–308 (2010)
7. Rossi, F., Liveri, M.L.T.: Chemical self-organization in self-assembling biomimetic systems. Ecol. Model. **220**(16), 1857–1864 (2009)
8. Sciascia, L., Rossi, F., Sbriziolo, C., Liveri, M.L.T., Varsalona, R.: Oscillatory dynamics of the Belousov-Zhabotinsky system in the presence of a self-assembling nonionic polymer. Role of the reactants concentration. Phys. Chem. Chem. Phys. **12**(37), 11674–11682 (2010)
9. Rossi, F., Varsalona, R., Marchettini, N., Turco Liveri, M.L.: Control of spontaneous spiral formation in a zwitterionic micellar medium. Soft Matter **7**, 9498 (2011)
10. Vanag, V.K., Epstein, I.R.: Pattern formation in a tunable medium: The Belousov-Zhabotinsky reaction in an aerosol OT microemulsion. Phys. Rev. Lett. **87**(22), 228301–4 (2001)
11. Toiya, M., Vanag, V.K., Epstein, I.R.: Diffusively coupled chemical oscillators in a microfluidic assembly. Angew. Chem. Int. Ed. **47**(40), 7753–7755 (2008)
12. Rossi, F., Vanag, V.K., Epstein, I.R.: Pentanary cross-diffusion in water-in—oil microemulsions loaded with two components of the Belousov-Zhabotinsky reaction. Chem. Eur. J. **17** (7), 2138–2145 (2011)
13. Tompkins, N., Li, N., Girabawe, C., Heymann, M., Ermentrout, G.B., Epstein, I.R., Fraden, S.: Testing turing's theory of morphogenesis in chemical cells. Proc. Natl. Acad. Sci. **111**(12), 4397–4402 (2014)
14. Walde, P., Umakoshi, H., Stano, P., Mavelli, F.: Emergent properties arising from the assembly of amphiphiles. Artificial vesicle membranes as reaction promoters and regulators. Chem. Commun. **50**(71), 10177–10197 (2014)
15. Tomasi, R., Noel, J.M., Zenati, A., Ristori, S., Rossi, F., Cabuil, V., Kanoufi, F., Abou-Hassan, A.: Chemical communication between liposomes encapsulating a chemical oscillatory reaction. Chem. Sci. **5**(5), 1854–1859 (2014)
16. Rossi, F., Zenati, A., Ristori, S., Noel, J.M., Cabuil, V., Kanoufi, F., Abou-Hassan, A.: Activatory coupling among oscillating droplets produced in microfluidic based devices. Int. J. Unconventional Comput. **11**(1), 23–36 (2015)

17. Torbensen, K., Rossi, F., Pantani, O.L., Ristori, S., Abou-Hassan, A.: Interaction of the Belousov-Zhabotinsky reaction with phospholipid engineered membranes. J. Phys. Chem. B **119**(32), 10224–10230 (2015)
18. Stockmann, T.J., Noël, J.M., Ristori, S., Combellas, C., Abou-Hassan, A., Rossi, F., Kanoufi, F.: Scanning electrochemical microscopy of Belousov-Zhabotinsky reaction: how confined oscillations reveal short lived radicals and auto-catalytic species. Anal. Chem. **87**(19), 9621–9630 (2015)
19. Taylor, A.F., Tinsley, M.R.,Wang, F., Huang, Z., Showalter, K.: Dynamical quorum sensing and synchronization in large populations of chemical oscillators. Science **323**(5914), 614–617 (2009)
20. Rossi, F., Ristori, S., Marchettini, N., Pantani, O.L.: Functionalized clay microparticles as catalysts for chemical oscillators. J. Phys. Chem. C **118**(42), 24389–24396 (2014)
21. Hu, G., Pojman, J.A., Scott, S.K., Wrobel, M.M., Taylor, A.F.: Base-catalyzed feedback in the urea-urease reaction. J. Phys. Chem. B **114**(44), 14059–14063 (2010)
22. Wrobel, M.M., Bánsági, T., Scott, S.K., Taylor, A.F., Bounds, C.O., Carranza, A., Pojman, J. A.: pH wave-front propagation in the urea-urease reaction. Biophys. J. **103**(3), 610–615 (2012)
23. Miele, Y., Bánsági, T., Taylor, A.F., Stano, P., Rossi, F.: Engineering enzyme-driven dynamic behaviour in lipid vesicles. In Rossi, F., Mavelli, F., Stano, P., Caivano, D. (eds.): Advances in artificial life, evolutionary computation and systems chemistry. Number 587 in communications in computer and information science, pp. 197–208. Springer International Publishing (2015)
24. Stingl, K., Altendorf, K., Bakker, E.P.: Acid survival of Helicobacter pylori: how does urease activity trigger cytoplasmic pH homeostasis? Trends Microbiol. **10**(2), 70–74 (2002)
25. Muzika, F., Bansagi, T., Schreiber, I., SchreiberovAą, L., Taylor, A.F.: A bistable switch in pH in urease-loaded alginate beads. Chem. Commun. (Cambridge, England) **50**(76), 11107–11109 (2014)
26. Lasic, D.D., Barenholz, Y.: Handbook of nonmedical applications of liposomes: theory and basic sciences, vol. 1. CRC Press (1996)
27. Paula, S., Volkov, A., Van Hoek, A., Haines, T., Deamer, D.W.: Permeation of protons, potassium ions, and small polar molecules through phospholipid bilayers as a function of membrane thickness. Biophys. J **70**(1), 339 (1996)
28. Pautot, S., Frisken, B.J., Weitz, D.A.: Production of unilamellar vesicles using an inverted emulsion. Langmuir **19**(7), 2870–2879 (2003)
29. Carrara, P., Stano, P., Luisi, P.L.: Giant vesicles colonies: a model for primitive cell communities. ChemBioChem **13**(10), 1497–1502 (2012)
30. Stano, P., Wodlei, F., Carrara, P., Ristori, S., Marchettini, N., Rossi, F.: Approaches to molecular communication between synthetic compartments based on encapsulated chemical oscillators. In Pizzuti, C., Spezzano, G., (eds.): Advances in Artificial Life and Evolutionary Computation. Number 445 in Communications in Computer and Information Science, pp. 58–74. Springer International Publishing (2014)
31. Ermentrout, B.: Simulating, analyzing, and animating dynamical systems: a guide to XPPAUT for researchers and students, vol. 14. Siam (2002)
32. Mathai, J.C., Sprott, G.D., Zeidel, M.L.: molecular mechanisms of water and solute transport across archaebacterial lipid membranes. J. Biol. Chem. **276**(29), 27266–27271 (2001)

Dissipative Particle Dynamics Study of Alginate/Gelatin Aerogels Obtained by Supercritical Drying

Simona Concilio, Stefano Piotto, Lucia Sessa, Lucia Baldino, Stefano Cardea and Ernesto Reverchon

Abstract The properties of alginate/gelatin (A/G) interpenetrated polymer networks have been studied by dissipative particle dynamics (DPD) simulations. The simulation predicted some mechanical properties of A/G blends with different A/G ratios in water. Results from new synthesized aerogels have been used to validate the range of exploitation of the DPD simulations. Good mechanical and morphological properties of the aerogels have been achieved from aerogels derived from hydrogels with water content higher than 95%. DPD simulation results indicated that an optimal shear viscosity is reached for a composition of 95% water, 3% alginate and 2% gelatin. Furthermore, this approach can be of great interest in designing novel materials.

Keywords DPD · Hydrogels · Supercritical drying

1 Introduction

Natural polymers, such as alginate and/or gelatin, can be used to produce scaffolds for tissue engineering applications, even though their mechanical and biochemical performance need often to be improved. A way to overcome these limitations is the generation of multi-component scaffolds, by blending two or more polymers. One way to realize this is the formation of an interpenetrating polymer network (IPN) between biopolymers, or the functionalization of one polymer with peptide sequences, in order to promote cell recognition [1].

S. Concilio (✉) · L. Baldino · S. Cardea · E. Reverchon
Department of Industrial Engineering, University of Salerno, Via Giovanni Paolo II 132, 84084 Fisciano, SA, Italy
e-mail: sconcilio@unisa.it

S. Piotto (✉) · L. Sessa
Department of Pharmacy, University of Salerno, Via Giovanni Paolo II 132, 84084 Fisciano, SA, Italy
e-mail: piotto@unisa.it

© Springer International Publishing AG 2018
S. Piotto et al. (eds.), *Advances in Bionanomaterials*,
Lecture Notes in Bioengineering, DOI 10.1007/978-3-319-62027-5_7

Alginate is largely used in biomedical field, due to its biodegradability, biocompatibility, hydrophilicity and low toxicity [2]. Nevertheless, its negative charge inhibits protein adsorption and reduces cellular adhesion. For this reason, bioactive molecules such as arginine-glycine-aspartic acid (RGD) and fibronectin are often proposed for the immobilization within the hydrogel, to induce cells adhesion [3–5]. Gelatin is formed by denatured collagen; it has relatively low antigenicity compared to its precursor and maintains signals that may promote cell adhesion, differentiation and proliferation, such as RGD sequence of collagen [6]. Interestingly, gelatin structure can be easily functionalized with anti-inflammatory compounds [7], cell membrane modulators [8, 9], or antimicrobial compounds [10, 11] for tissue engineering applications.

Alginate and gelatin are often mixed and prepared in form of hydrogels for soft tissue applications, or dried by lyophilization, for hard tissues application. In the latter case, the scaffold can collapse, due to the surface tension exerted by the solvent on the polymer matrix during the drying process. The preparation of biocompatible scaffold for tissue engineering requires good mechanical properties and high porosity. Nevertheless, scaffolds are generally characterized by poor mechanical properties and could not have the correct porosity to let the cells interact to each other during the recognition process. For example, when freeze drying is used for biopolymer aerogel production, the authors often obtain microporous aerogels that do not exhibit a sub-nanostructure, but an irregular and closed morphology [12–14].

The extraction of solvent from the hydrogel is a key moment because the inner fluid must be removed without the collapse of the structure. The decline of fluid pressure in connection with the withdrawal of fluid from the hydrogel gives rise to a change in volume of both fluids and gel structure. The volume variation of the matrix results in a decrease of both the pore volume and the total volume of the fluid-filled formation.

In our previous works, interpenetrated alginate/gelatin hydrogels have been successfully obtained and preserved by supercritical carbon dioxide drying (SC-CO$_2$), performed at 200 bar and 35 °C [15]. The process allowed modulation of morphology and mechanical properties of these blends, and the complete elimination of crosslinking agents, such as glutaraldehyde [16]. It was possible to obtain this result thanks to a supercritical drying process that, working at zero surface tension, allows preserving the hydrogels nanostructure in the corresponding aerogels.

Hydrogels with high shear viscosity are those that can successfully resist to the drying process. The shear and bulk viscosities of a potential scaffold are important properties, in relation with the final use of the system. More specifically, the translational molecular motion is associated with the shear viscosity. The resistance of a fluid to compression is related to the bulk viscosity.

In this work, we explored the possibility to use coarse grain simulations, namely dissipative particle dynamics (DPD), for calculating shear and bulk viscosity in ternary systems of water, alginate and gelatin. The present work consisted in running molecular dynamics for calculating the Flory-Huggins parameters needed for

DPD calculations. Flory-Huggins parameters permitted to run DPD calculations for 15 ternary systems with water content higher than 95%. Moreover, some new A/G aerogels obtained with a series of different polymer composition were prepared, in order to explore and validate the computational results.

A rationalization of experimental results can be obtained from DPD simulation, which allows to predict and explain the different morphologies we obtain experimentally. As a result of this study, it will be possible to predict and select the final scaffold structure by changing the polymers composition.

2   Materials and Methods

2.1   Dissipative Particle Dynamics

The simulation of the aggregation of polymer mixture at different composition was carried out employing an efficient coarse grain algorithm, i.e. the dissipative particle dynamics (DPD). The DPD approach is a powerful, off-lattice, dynamical model. In a DPD simulation, the number of particles are significantly lower than a full atomistic representation and each particle represents the center of mass of a cluster of atoms.

The simplest and best-known theory of the thermodynamics of mixing and phase separation in binary systems is the Flory-Huggins model [7]. The interaction parameter, χ, is defined as:

$$\chi = \frac{E_{mix}}{RT}$$

where E_{mix} is the mixing energy; that is, the difference in free energy due to interaction between the mixed and the pure state.

The coarse graining of all particles of this study was performed following the original protocol of Groot and Warren [17]. DPD repulsive parameters are linearly correlated to the mixing energies. In the present work, we used the CVFF force field, [18–20] that reproduce correctly the experimental observations. The relationship between the Flory-Huggins parameter and the repulsion parameter is:

$$a_{ij} = 25 + 3.50\chi$$

Because most of the parameterization of polymers were made in aqueous solutions, we calculated the free energy of mixing among all the species listed in Table 1 and water clusters. The values of χ were rescaled to completely adhere to the original Groot-Warren interaction values. The peptides have been parameterized according to Neimark and coworkers [21].

Table 1 DPD repulsive parameters obtained via molecular dynamics protocol [19] and from the work of Vishnyakov [21] for the peptide intrachain terms (N and P)

	Glycine	N	P	Water	Alginate	Hyd proline	Proline	Alginate node	Gelatin node
Glycine	40								
N	35	80							
P	35	12	80						
Water	40	25	25	25					
Alginate monomer	50	20	20	25	25				
Hyd proline	40	30	30	30	25	32			
Proline	40	50	50	60	40	40	30		
Alginate node	50	20	20	25	25	40	25	10	
Gelatine node	40	35	35	40	50	40	40	50	10

For the simulation of alginate and gelatin, several types of particles were used and their mutual interactions were expressed in terms of repulsion parameters. The DPD parameters are related to the compressibility of fluid and to Flory-Huggins solubility parameters such that a reasonable description of the thermo-dynamics of the real system can be obtained.

The topology of alginate consists in a string of 50 beads with two random positions for cross-linking. The sodium ion was not explicitly considered and was included in the water simulation. Gelatin was represented of 63 beads (the triplet Gly-Pro-Hyp was repeated 21 times), with two random positions for cross-linking. The link between alginate polymers and between gelatin polymers was reproduced with a repulsive parameter set to 10.

2.2 Mechanical Properties Calculations

Application of the Green-Kubo formalism to the shear viscosity gives the following relation between shear viscosity and the time correlation functions involving the off diagonal components of the pressure tensor:

$$\eta = \frac{V}{kT} \int_0^\infty \langle P_{\alpha\beta}(t)P_{\alpha\beta}(0)\rangle dt$$

where the $P_{\alpha\beta}$ denote the three equivalent off diagonal elements of the (instantaneous) pressure tensor.

The bulk viscosity η_v is related to the decay of fluctuations in the diagonal elements of the stress tensor as follows:

$$\eta_v = \frac{V}{kT} \int_0^\infty \langle \delta P(t) \delta P(0) \rangle dt$$

where $\delta P = P = P - <P>$.

2.3 Production of Alginate/Gelatin Aerogels

PROTANAL LF 10/60 was termed high-G alginate, with M/G = 25/75 and was kindly provided by FMC BioPolymer; gelatin, type B from bovine skin, Calcium Chloride, Glutaraldehyde solution 25% w/w in water and Ethanol (purity > 99.8%) were purchased from Sigma-Aldrich; Carbon Dioxide (99% purity) was bought from Morlando Group S.R.L. (Sant'Antimo, NA—Italy). Distilled water was produced using a laboratory water distiller supplied by ISECO (St. Marcel, AO—Italy). All materials were used as received.

Experimental procedure to produce alginate/gelatin aerogel

Two aqueous solutions of alginate (2% w/w) and gelatin (2% w/w) were prepared, as described in ref [15]. The solutions were stirred for 24 h at 200 rpm. Then, the solutions of alginate and gelatin were mixed in three different ratios by volume: A/G—20/80, 50/50, 80/20. The mixtures were stirred for 1 h and, then, poured into cylindrical molds of 2 cm diameter and thickness of about 2 mm. Each sample was immersed in 25 mL of $CaCl_2$ at 5% w/w in water for 24 h, to promote Alginate gel formation. The obtained hydrogels were repeatedly washed with distilled water to remove Ca^{2+} residues. Afterwards, samples were immersed in 30 mL of an aqueous solution of GTA (aqueous GTA, 25% w/w) 8% v/v to induce the crosslinking of gelatin. A/G hydrogels were rinsed three times in distilled water to remove GTA excess and underwent to the same process of multistage exchange of the solvent used for single polymer hydrogels.

Supercritical gel drying

A/G aerogels were prepared using a homemade laboratory plant that consists of a 316 stainless steel cylindrical high-pressure vessel with an internal volume of 200 mL, equipped with a high pressure pump (mod. LDB1, Lewa, Germany) used to deliver SC-CO_2. Pressure in the vessel was measured by a test gauge (mod. MP1, OMET, Italy) and regulated using a micrometering valve (mod. 1335G4Y, Hoke, SC, USA). Temperature was regulated using PID controllers (mod. 305, Watlow, USA). At the exit of the vessel, a rotameter (mod. D6, ASA, Italy) was used to measure CO_2 flow rate. The vessel was filled with SC-CO_2; then, when the required pressure and temperature were obtained (200 bar and 35 °C), drying was performed using a SC-CO$_2$ flow rate of about 1 kg/h for 5 or 8 h. A depressurization time of about 30 min was used to bring back the system at atmospheric pressure.

3 Results

Aerogels formed by blends of alginate (A) and gelatin (G), were observed by FESEM to analyze their structure. In Fig. 1, pictures of W/A/G aerogel at different percentages are reported, in order to evaluate the changes in morphologies due to the different polymer composition. Gelatin aerogels, W/A/G 95-0-5 (Fig. 1a), were characterized by a nanofibrous structure, while Alginate aerogels W/A/G 95-5-0 (Fig. 1b) showed a nanoporous homogeneous structure. W/A/G aerogels morphology changed with the relative proportion of the two polymers; the trend was less evident for W/A/G 98-0.4-1.6, 98-1-1 and 98-1.6-0.4, showing that an optimal

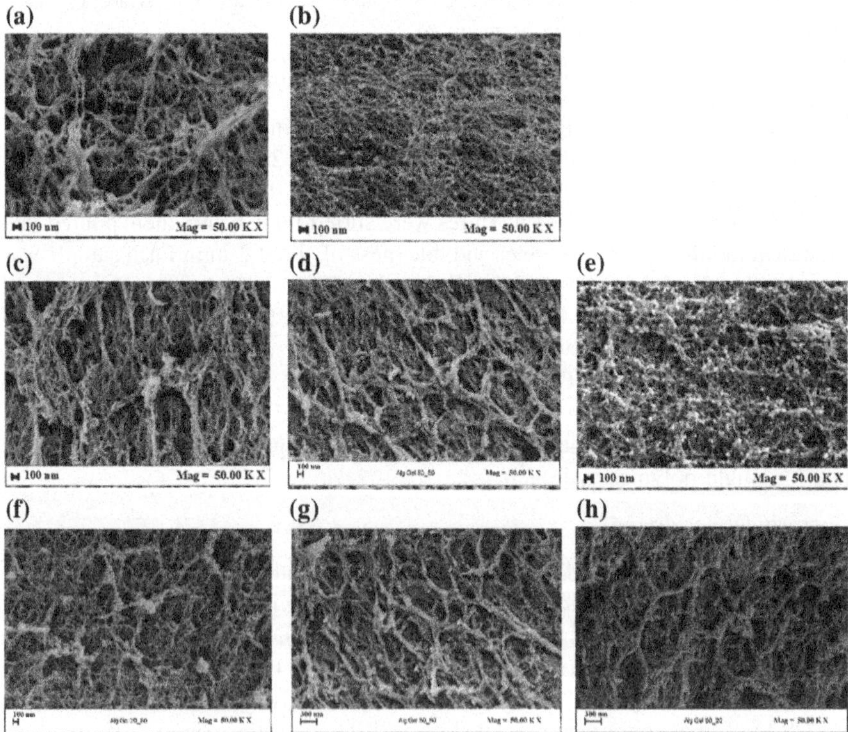

Fig. 1 Morphology of A/G blend aerogels: **a** W/A/G 95-0-5, **b** W/A/G 95-5-0, **c** W/A/G 95-1-4, **d** W/A/G 95-2.5-2.5, **e** W/A/G 95-4-1, **f** W/A/G 98-0.4-1.6, **g** W/A/G 98-1-1, **h** W/A/G 98-1.6-0.4. Magnification 50,000 X. Pictures **a–e** are taken from Ref. [15]

Table 2 Comparison among tensile mechanical properties of W/A/G aerogels, processed at 200 bar, 35 °C, 5 h

W/A/G Aerogel	Young modulus [MPa]	Tensile strength at break [MPa]
95-0-5[a]	0.91 ± 0.11	1.41 ± 0.15
95-1-4[a]	0.85 ± 0.08	1.92 ± 0.18
95-2.5-2.5[a]	0.78 ± 0.06	2.33 ± 0.25
95-4-1[a]	0.61 ± 0.05	2.54 ± 0.30
95-5-0[a]	0.48 ± 0.03	2.78 ± 0.36
98-0.4-1.6	0.47 ± 0.07	0.74 ± 0.04
98-1-1	0.39 ± 0.04	1.09 ± 0.06
98-1.6-0.4	0.27 ± 0.04	1.35 ± 0.09

[a]Values from Ref. [15]

morphology distribution can be obtained with at least 5% of polymers in water. The water content cannot be higher than 95%, in order to favor a proper polymer crosslink and appreciate the hydrogel morphology changes. As a result, it is possible to select the scaffold structure by changing the polymers relative proportions. In all cases, the nanoscale morphology was preserved. The supercritical process does not modify gel organization and avoid of gel collapse during drying of A/G blends.

In Table 2, tensile mechanical properties of W/A/G aerogels are reported.

The aerogels of single polymers (W/A/G 95-0-5 and 95-5-0) are characterized by a higher Young modulus for gelatin (0.91 MPa) and a higher tensile strength at break for alginate (2.78 MPa). Combining the two polymers, intermediate values are obtained. This was also confirmed for the series W 98-0.4-0.6 to 98-1.5-0.4, but in these cases the values are considerably lower that the series with 95% of water, indicating that in this case the polymer content is too low for the formation of a solid network. These results confirm that it is possible to continuously modulate A/G aerogels mechanical properties, since gelatin produces an increase of the elasticity of alginate, and alginate increases the tensile strength of gelatin, producing an aerogel with intermediate characteristics.

3.1 DPD Results

A wider range of W/A/G aerogel compositions were analyzed by DPD, from 95 to 98% water content, with at different A/G ratios blends. We have performed 15 DPD simulations at compositions listed in Table 3.

We explored the balance of alginate and gelatin to promote cross-linking still preserving the viscosity that is need for safe solvent removal. The results can be plotted in a ternary diagram. The viscosity values are colormapped.

Unsurprisingly, the bulk viscosity has a maximum for pure water system and decreases monotonically along with the water content. Contrarily, the more

Table 3 DPD simulations of bulk and shear viscosity for different W/A/G compositions. The errors are lower than 5%

W/A/G Aerogel	Bulk viscosity 10^8 (Pa·s)	Shear viscosity (Pa·s)
95	1.27	5.15
95-4-1	1.30	21.4
95-3-2	1.32	77.3
95-2.5-2.5	1.34	26.2
95-2-3	1.35	21.1
95-1-4	1.38	22.8
95-0-5	1.40	24.8
96-0-4	1.42	1.61
96-4-0	1.30	28.0
96-1-3	1.39	49.0
96-3-1	1.33	21.7
98-1-1	1.41	4.08
98-0.4-1.6	1.42	3.22
98-1.6-0-4	1.39	4.98
100-0-0	1.45	5.06

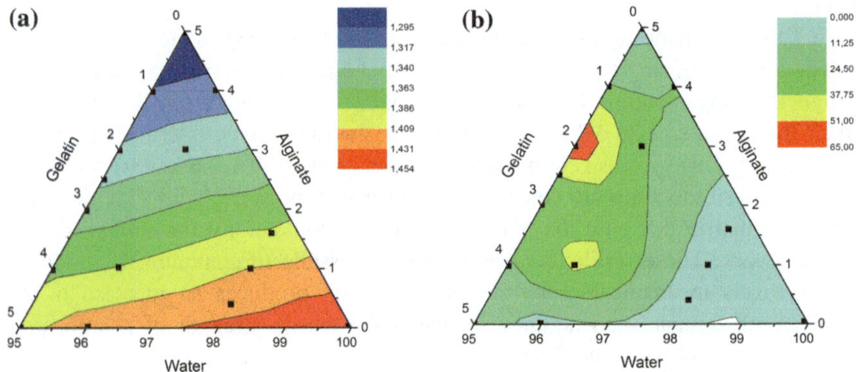

Fig. 2 Ternary diagrams for the system water-alginate-gelatin with bulk viscosity (**a**) and shear viscosity (**b**)

important parameter, the shear viscosity, has a more complex behavior. As shown in Fig. 2b the maximum of shear viscosity is reached for a system made of 95% water, 3% alginate and 2% gelatin. Though the red region can be considered optimal, also green region shows a shear viscosity above 30 Pa·s that is sufficient, under mild and controlled extraction, to prevent the collapse of the network.

DPD results confirm the experimental observation that a minimum water content of 95% is necessary in order to obtain good mechanical properties of the scaffold.

4 Conclusion

The preparation of a proper scaffold for tissue engineering is a complex procedure that must consider, among other parameters, the correct porosity, biocompatibility, high level of interconnection and good mechanical properties. Here we concentrated on exhaustive investigation of the hydrogel preparation.

The most important parameter for a hydrogel is the shear viscosity that must be high enough to prevent the collapse of the three dimensional network during the SC-CO_2 extraction. We used DPD to calculate shear and bulk viscosity in 15 ternary systems with water content equal or higher than 95%. The maximum in shear viscosity is reached for a composition of 95% water, 3% alginate and 2% gelatin. The region between 97 and 95% of water show acceptable shear viscosity, in order to prevent the collapse of the network. The water content in any case cannot be higher than 98%. We synthesized new samples that confirm the DPD results. Taken together, these results indicate a reliable strategy for the design of biomaterials, using MD and DPD simulations. The study indicates that the combination of experiments and the computer simulation can be an effective way of designing and developing the polymer blends with the tailored properties.

References

1. Autissier, A., Le Visage, C., Pouzet, C., Chaubet, F., Letourneur, D.: Fabrication of porous polysaccharide-based scaffolds using a combined freeze-drying/cross-linking process. Acta Biomater. **6**(9), 3640–3648 (2010)
2. Jawad, H., Lyon, A.R., Harding, S.E., Ali, N.N., Boccaccini, A.R.: Myocardial tissue engineering. Br. Med. Bull. **87**(1), 31–47 (2008)
3. Baldwin, A.D., Kiick, K.L.: Polysaccharide-modified synthetic polymeric biomaterials. Pept. Sci. **94**(1), 128–140 (2010)
4. Freeman, I., Cohen, S.: The influence of the sequential delivery of angiogenic factors from affinity-binding alginate scaffolds on vascularization. Biomaterials **30**(11), 2122–2131 (2009)
5. Wu, X., Liu, Y., Li, X., Wen, P., Zhang, Y., Long, Y., Wang, X., Guo, Y., Xing, F., Gao, J.: Preparation of aligned porous gelatin scaffolds by unidirectional freeze-drying method. Acta Biomater. **6**(3), 1167–1177 (2010)
6. Inzana, J.A., Olvera, D., Fuller, S.M., Kelly, J.P., Graeve, O.A., Schwarz, E.M., Kates, S.L., Awad, H.A.: 3D printing of composite calcium phosphate and collagen scaffolds for bone regeneration. Biomaterials **35**(13), 4026–4034 (2014)
7. Lopez, D.H., Fiol-deRoque, M.A., Noguera-Salva, M.A., Teres, S., Campana, F., Piotto, S., Castro, J.A., Mohaibes, R.J., Escriba, P.V., Busquets, X.: 2-hydroxy arachidonic acid: a new non-steroidal anti-inflammatory drug. PLoS ONE **8**(8), e72052 (2013)
8. Piotto, S., Concilio, S., Bianchino, E., Iannelli, P., Lopez, D.J., Teres, S., Ibarguren, M., Barcelo-Coblijn, G., Martin, M.L., Guardiola-Serrano, F., Alonso-Sande, M., Funari, S.S., Busquets, X., Escriba, P.V.: Differential effect of 2-hydroxyoleic acid enantiomers on protein (sphingomyelin synthase) and lipid (membrane) targets. Biochem. Biophys. Acta. **1838**(6), 1628–1637 (2014)

9. Piotto, S., Trapani, A., Bianchino, E., Ibarguren, M., Lopez, D.J., Busquets, X., Concilio, S.: The effect of hydroxylated fatty acid-containing phospholipids in the remodeling of lipid membranes. Biochem. Biophys. Acta. **1838**(6), 1509–1517 (2014)
10. Piotto, S., Concilio, S., Sessa, L., Porta, A., Calabrese, E.C., Zanfardino, A., Varcamonti, M., Iannelli, P.: Small azobenzene derivatives active against bacteria and fungi. Eur. J. Med. Chem. **68**(178–184 (2013)
11. Piotto, S., Concilio, S., Sessa, L., Iannelli, P., Porta, A., Calabrese, E.C., Galdi, M.R., Incarnato, L.: Novel antimicrobial polymer films active against bacteria and fungi. Polym. Compos. **34**(9), 1489–1492 (2013)
12. Cheng, Y., Lu, L., Zhang, W., Shi, J., Cao, Y.: Reinforced low density alginate-based aerogels: preparation, hydrophobic modification and characterization. Carbohyd. Polym. **88**(3), 1093–1099 (2012)
13. Yamamoto, M., James, D., Li, H., Butler, J., Rafii, S., Rabbany, S.: Generation of stable co-cultures of vascular cells in a honeycomb alginate scaffold. Tissue Eng. Part A **16**(1), 299–308 (2009)
14. Zhang, F., He, C., Cao, L., Feng, W., Wang, H., Mo, X., Wang, J.: Fabrication of gelatin—hyaluronic acid hybrid scaffolds with tunable porous structures for soft tissue engineering. Int. J. Biol. Macromol. **48**(3), 474–481 (2011)
15. Baldino, L., Concilio, S., Cardea, S., Reverchon, E.: Interpenetration of natural polymer aerogels by supercritical drying. Polymers **8**(4), 106 (2016)
16. Baldino, L., Concilio, S., Cardea, S., De Marco, I., Reverchon, E.: Complete glutaraldehyde elimination during chitosan hydrogel drying by SC-CO$_2$ processing. J. Supercrit. Fluids **103**(70–76 (2015)
17. Groot, R.D., Warren, P.B.: Dissipative particle dynamics: bridging the gap between atomistic and mesoscopic simulation. J. Chem. Phys. **107**(4423–4435 (1997)
18. Bianchino, E., Piotto, S., Mavelli, F., Curri, M.L., Striccoli, M.: DPD simulations of PMMA-Oleic acid mixture behaviour in organic capped nanoparticle based polymer nanocomposite. Macromol. Symp. **286**(1), 156–163 (2009)
19. Caracciolo, G., Piotto, S., Bombelli, C., Caminiti, R., Mancini, G.: Segregation and phase transition in mixed lipid films. Langmuir **21**(20), 9137–9142 (2005)
20. Amenitsch, H., Bombelli, C., Borocci, S., Caminiti, R., Ceccacci, F., Concilio, S., La Mesa, C., Mancini, G., Piotto, S., Rappolt, M.: Segregation into domains observed in liquid crystal phases: comparison of experimental and theoretical data. Soft Matter **7**(7), 3392–3403 (2011)
21. Vishnyakov, A., Talaga, D.S., Neimark, A.V.: DPD simulation of protein conformations: from alpha-Helices to beta-structures. J. Phys. Chem. Lett. **3**(21), 3081–3087 (2012)

Molecular Dynamics and Morphing Protocols for High Accuracy Molecular Docking

Lucia Sessa, Simona Concilio and Stefano Piotto

Abstract Molecular docking is a popular technique to analyse the geometry and the interactions of a ligand in a protein binding site. Flexibility in molecular docking studies is particularly important when the binding pocket is buried inside the protein and the ligand binding is responsible for backbone deformation of the receptor. The major limit is that the most popular docking programs do not consider the conformation changes in the protein of the receptor during the process of binding of the ligand. Here we have considered sampling of molecular dynamics trajectory, and morphing protocol to generate conformers of the receptor, which differs from the available crystal structure. We have also considered the presence of conserved residues to drive ligands toward the binding pocket under blind docking analysis.

Keywords Flexible docking · Yada · Cross-docking

1 Introduction

The main goal in drug discovery is the accurate evaluation of ligand binding to a protein receptor. The computational techniques are gaining interest because they reduce the time and the cost of experimental tests to identify a bioactive compound. In this field, molecular docking is the most used tool to analyse the geometry and the interactions of a ligand in a protein binding site. Recently [1, 2], docking technique has been used successfully for designing new drugs with high binding affinity [3].

L. Sessa (✉) · S. Piotto
Department of Pharmacy, University of Salerno, Via Giovanni Paolo II 132,
84084 Fisciano, SA, Italy
e-mail: lucsessa@unisa.it

S. Concilio
Department of Industrial Engineering, University of Salerno, Via Giovanni Paolo II 132,
84084 Fisciano, SA, Italy

© Springer International Publishing AG 2018
S. Piotto et al. (eds.), *Advances in Bionanomaterials*,
Lecture Notes in Bioengineering, DOI 10.1007/978-3-319-62027-5_8

The aim of this study is to explore the limits of current docking method and the best strategies to improve the accuracy. The major problem is that the most popular docking programs do not consider the conformation changes in the protein of the receptor during the process of binding of the ligand. When the three-dimensional structure of a protein is known, it only represents the protein conformation under the condition of the X-ray crystallography experiment. Crystallographic data show that different ligands may stabilize different receptor conformations. Consequently, the 3D structure represents only a successful binding event of a specific protein conformation and ligand. Because a single representative structure for the receptor in docking study is not enough, the flexible docking is widely used today. Flexibility in molecular docking studies is particularly important when the binding pocket is buried inside the protein and the ligand binding induces the backbone deformation of the receptor to permit the ligand to get in. Cross-docking is a validation procedure consisting in docking a series of ligands into different conformations of the same receptor and calculating root mean square deviations (RMSD) between the predicted structure of a ligand/receptor complex with the better binding energy and the crystallographic complex conformation. Since the structures of the same receptor can be rather different, the cross-docking results are typically very poor [4]. It is well known [5–8] that proteins acquire different conformations upon binding with ligands.

Typically, receptor flexibility, when considered, is limited to side chain mobility. This flexibility is easy to perform and is available on the most popular docking software. Unfortunately, with few exceptions represented by surface binding sites, it does not offer a general and reliable procedure and it fails when the ligands are confined in the interior of the protein. Recently, and similarly to the work of Zhao et al. [9], we started to explore the possibility of using molecular dynamics (MD) to generate alternative backbone conformations. MD is an inexpensive method to generate a series of target structures for docking analysis. MD deforms the shape of the cavity that may now accommodate the ligand in the docking studies. Unfortunately, based on the docking results the MD provided various conformations with different internal cavities but this was not enough to improve the docking results. The problem of docking can be reduced to the search of the best method to add backbone flexibility without avoidable deformation. In the present work, we have considered sampling of molecular dynamics trajectory, and morphing protocol to generate conformers of the protein, which differs from the available crystal structure. We have also considered the presence of conserved residues to drive ligands toward the binding pocket under blind docking analysis [10].

The use of a single structure of the receptor can radically affect the outcome and alter the cross-docking results. This is because in that conformation the buried binding pocket is not large enough to accommodate all ligands. In these cases, the ligands remain confined on the receptor molecular surface. Here we take into account the androgen receptor (AR) since several receptor/ligand structures are known, and it provides a good set for docking method validation.

2   Materials and Methods

2.1   Data Set

The androgen receptor is a nuclear receptor that controls the expression of specific genes [11] with a binding site well buried inside. The natural ligands are 5α-dihydrotestosterone (DHT) and testosterone (TES) [12]. In the PDB database there are 204 structures collected *via* X-ray crystallography and nuclear magnetic resonance (NMR). In our study, we selected 11 crystallographic structures of AR in complex with 11 different ligands (see Table 1). The proteins showed the 100% of sequence identity. To estimate the conformational variability of the experimental structure of AR, we report the RMSD values calculated between the receptor structure in 3L3X (the crystallographic structure with the native ligand) and the receptor structure in the other 10 AR conformation. The resolution of the protein/ligand complex was smaller than 2 Å. The binding site is buried inside the protein; it represents a cavity that hosts the ligand. The position of the binding site was the same in all structures, while the shape and the volume are rather different. We calculated, for all receptors, the Van der Waals surface using 1.4 Å as radius of the water probe. In Table 1 the volumes of the cavities and of the respective ligands are compared. For two receptors (2AM9 and 3B67) the binding pocket is not insulated from the solvent and, therefore, there are not cavities to calculate.

In Table 2 the structures of the 11 ligands are shown. The ligand R18 is the only ligand with a steroid structure similar to the natural ligand DHT and TES.

Table 1   Receptor dataset

PDB	Ligand ID	Resolution (Å)	RMSD respect to 3L3X (Å)	Volume cavity (A^3)	Volume ligand (A^3)
3L3X	DHT	1.55	0.000	561.47	322.62
2AM9	TES	1.64	0.469	n.a.	323.81
1XOW	R18	1.80	0.389	520.6	315.83
3B5R	B5R	1.80	0.572	650.8	395.87
3B65	3B6	1.80	0.602	651.5	375.26
3B66	B66	1.65	0.573	648.6	393.86
3B67	B67	1.90	0.512	n.a.	406.27
3B68	B68	1.90	0.606	767.30	420.89
2AX8	FHM	1.7	0.454	657.3	375.61
2AX9	BHM	1.65	0.464	555.6	295.86
2AX6	HFT	1.5	0.521	520.1	268.44

Table 2 Ligand dataset

PDB	ID	Name	Structure
3L3X	DHT	5-alpha-dihydrotestosterone	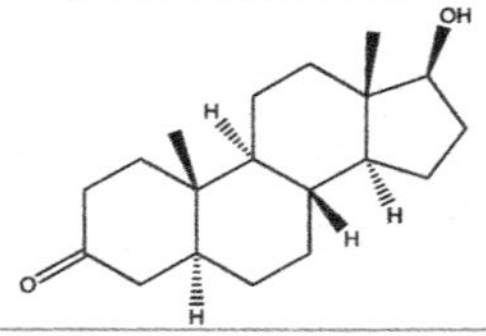
2AM9	TES	Testosterone	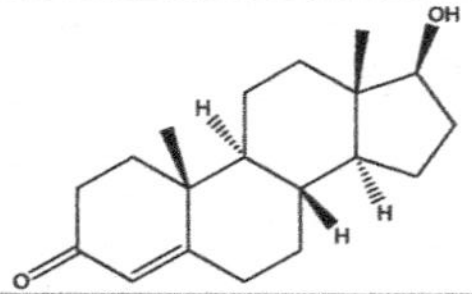
1XOW	R18	(17beta)-17-hydroxy-17-methylestra-4,9,11-trien-3-one	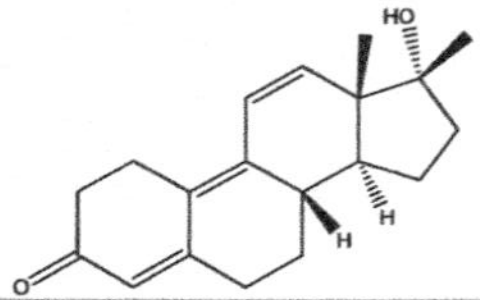
3B5R	B5R	(2S)-3-(4-chloro-3-fluorophenoxy)-N-[4-cyano-3(trifluoromethyl)phenyl]-2-hydroxy-2-methylpropanamide	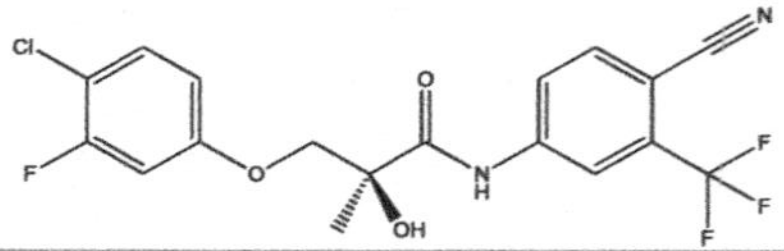
3B65	3B6	(2S)-N-(4-cyano-3-iodophenyl)-3-(4-cyanophenoxy)-2-hydroxy-2-methylpropanamide	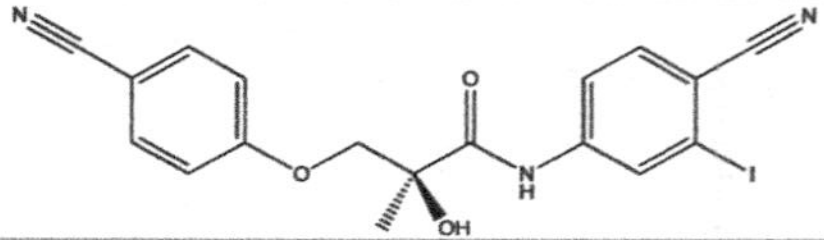

(continued)

Table 2 (continued)

PDB	ID	Name	Structure
3B66	B66	4-{[(1R,2S)-1,2-dihydroxy-2-methyl-3-(4-nitrophenoxy)propyl]amino}-2-(trifluoromethyl) benzonitrile	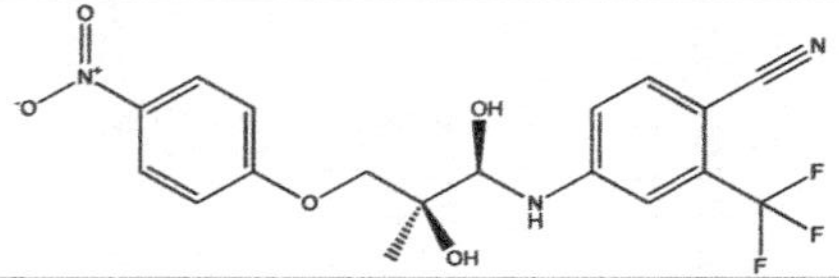
3B67	B67	(2S)-2-hydroxy-2-methyl-N-[4-nitro-3-(trifluoromethyl)phenyl]-3-(pentafluorophenoxy) propanamide	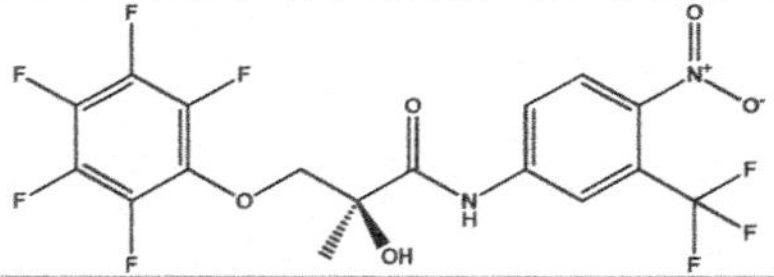
3B68	B68	(2S)-3-[4-(acetylamino)phenoxy]-2-hydroxy-2-methyl-N-[4-nitro-3-(trifluoromethyl)phenyl] propanamide	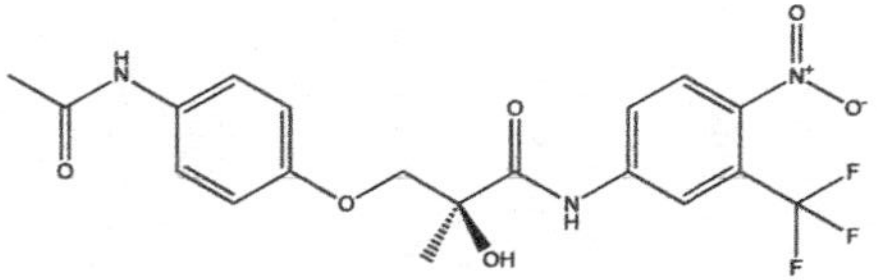
2AX8	FHM	S-3-(4-fluorophenoxy)-2-hydroxy-2-methyl- N-[4-nitro-3-(trifluoromethyl)phenyl] propanamide	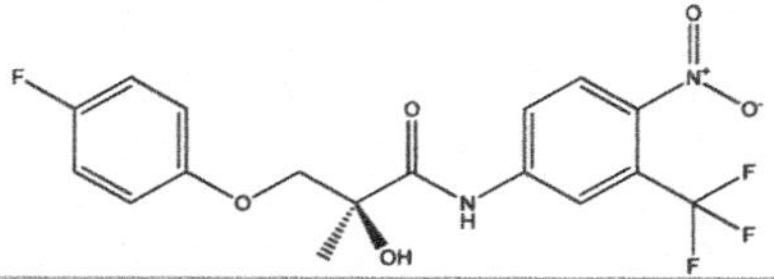
2AX9	BHM	(R)-3-bromo-2-hydroxy-2-methyl-N-[4-nitro-3-(trifluoromethyl)phenyl] propanamide	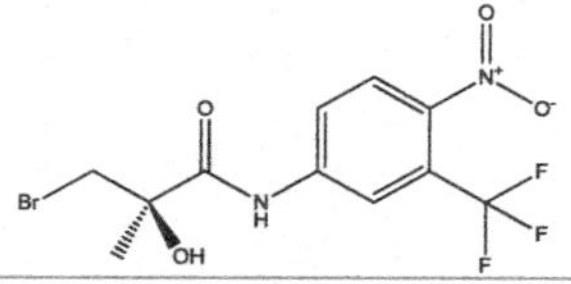
2AX6	HFT	2-hydroxy-2-methyl-N-(4-nitro-3-(trifluoromethyl)phenyl) propanamide	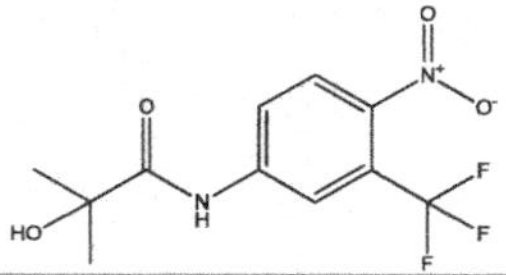

2.2 Sampling Stage

The flexible docking protocol [4] permits to explore the receptor conformational variability. Proteins extracted from crystal structure were prepared by adding hydrogen atoms with the protonation state simulated to pH = 7.4. All water molecules and molecules different form the ligand compound (such as glycerol etc.) were removed. A molecular dynamics simulations (MD) of the complexes were performed using a method already described in our previous work [4]. We collected from MD 5000 conformations for each ligand/receptor complex at regular time interval (each 500 ps).

2.3 Morphing

Morphing method generates a clash-free, physically rational, trajectory connecting two or more conformations of a macromolecule. We created a hybrid receptor structure (HYBR) transforming a structure with small cavity (3L3X with 561 $\mathring{A}^3$) into a structure with bigger cavity (3B68 with 767 $\mathring{A}^3$) during a simulated annealing minimization. A low-energy pathway between the two conformations was calculated with YASARA [13] using the Amber14 force field [14]. After removing clashes with an initial unrestrained steepest descent minimization, the procedure continued with a restrained simulated annealing minimization (time step 2 fs, atom velocities scaled down by 0.9 every 10th step). A force of 100 Piconewton (pointing towards the target coordinates) was applied to all the atoms and the RMSD from the target coordinates was measured every 10th step. If the RMSD improved less than 0.01 A, the pulling force was increased by 50%, till a maximum of 10,000 pN.

2.4 Docking Stage

The molecular docking simulations were performed using a new version of the software YADA [15] in which the docking analysis is correlated to the receptor sequence conservation. In addition, YADA exploits a novel genetic algorithm to improve the pose ranking. To reduce the computational time we used a specialized grid (GRIMD) to distribute the calculations [16].

The AR protein is highly conserved through species, the binding pocket (Fig. 1) is extremely conserved and, therefore, the use of YADA was particularly appropriate.

We build the conservation string in two steps. Using the web tool BLAST [17], we collected 1000 proteins reported in Uniprot database [18] with similar sequence

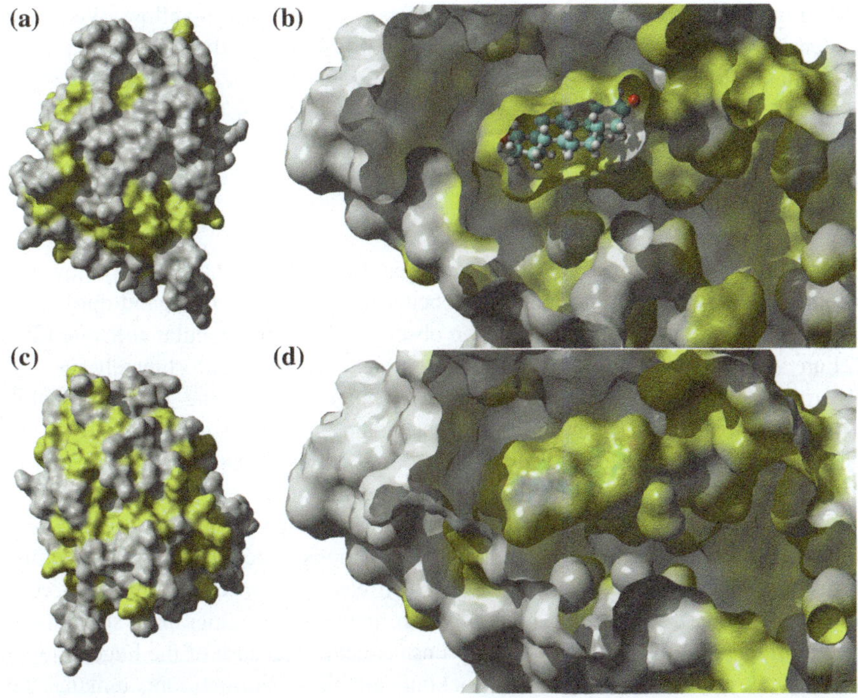

Fig. 1 Molecular surface of AR (PDB 3L3X). In yellow are the highly conserved residues of the protein, in grey are showed the residues with low conservation trough the species (**a** and **c**). Section of the conserved binding pocket housing the DHT (**b** and **d**)

to AR. Then the conserved regions were extracted by multiple sequence analysis using the tool online ClustalOmega [19].

The conservation string built for AR is the following:

Seq 1–50	SQPIFLNVLEAIEPGVVCAGHDNNQPDSFAALLSSLNELGERQLVHVVKW
Cons-weight	48815547953999478679495333984021895599295669761699
Seq 51–100	AKALPGFRNLHVDDQMAVIQYSWMGLMVFAMGWRSFTNVNSRMLYFAPDL
Cons-weight	69389999889767996468989992864968799993445537989999
Seq 101–150	VFNEYRMHKSRMYSQCVRMRHLSQEFGWLQITPQEFLCMKALLLFSIIPV
Cons-weight	87880893271674392184365479259865069977994888853892
Seq 151–200	DGLKNQKFFDELRMNYIKELDRIIACKRKNPTSCSRRFYQLTKLLDSVQP
Cons-weight	79974920979793499699262831324522225499889988996563
Seq 201–250	IARELHQFTFDLLIKSHMVSVDFPEMMAEIISVQVPKILSGKVKPIYFHT
Cons-weight	64349328482384353243969979839895196996349258664993

The residues with conservation weight > 7 were treated as conserved amino acid.

In Fig. 1 the conserved regions in AR protein are mapped in yellow. We put in evidence a section of the binding pocket (B and D) housing the DHT. The rest of the protein surface is represented in grey.

3   Results

The crystallographic structures of 11 AR/ligand complexes show clearly how the receptors deform their conformations to better accommodate the ligand during the binding. In some re-docking studies, we observed that some popular software [20–22] are very accurate in reproducing the complex structure. Unfortunately, in case of cross-docking, that is the docking of a ligand against the crystallographic structure of the same protein with different ligand, the results are extremely poor. There is a large variation of accuracy in cross-docking if ligands different in size, shape or chemical structure are docked into one structure.

Our goal was to improve the output introducing some degree of receptor flexibility. We collected different receptor conformations at regular intervals of time during a long MD simulation. The MD is an easy way to collect different conformations of the receptor that hopefully can match a crystallographic structure.

Unfortunately, this method is largely unsatisfactory because of the huge amount of data required for increasing the docking reliability. Furthermore, remains the problem to identify the correct receptor structure and to perform a docking analysis on that specific structure.

To validate the procedure of flexible docking, the predicted pose with the highest binding energy was superposed on the crystallographic complex conformation to calculate the root mean square deviation (RMSD).

The overall results of re-docking as listed in Table 3 is excellent (italic), but the cross-docking is severely limited.

The cross-docking study shows that small ligands like testosterone, are not docking properly in AR receptor with large binding cavities. Larger ligands such as 3B6, B67 and B68, 4 times out of 11 did not enter in the receptor.

The application of morphing permitted to generate a new geometry (HYBR) that greatly improved the quality of docking. Even large ligands (3B6 and B68) can enter the structure (Fig. 2a and b). R18 is the only ligand with a steroid structure like testosterone and DHT, with a low RMSD in all structures but with a RMSD of 3.75 Å in HYBR. This is because the binding pocket of HYBR is large enough to accommodate R18 in an upside-down conformation (Fig. 2c).

Table 3 RMSD of cross-docking calculations

	1XOW	2AM9	2AX6	2AX8	2AX9	3B5R	3B65	3B66	3B67	3B68	3L3X	HYBR
3B6	13.73	8.37	14.80	1.78	16.92	1.69	*1.63*	1.65	1.78	1.65	17.12	3.67
B5R	15.43	7.86	15.78	1.27	16.56	*0.99*	1.03	1.81	1.83	1.78	16.73	4.26
B66	16.22	7.57	15.83	1.83	17.09	0.97	0.80	*1.76*	1.77	1.67	17.26	3.95
B67	16.39	17.23	14.74	0.98	16.13	1.21	1.96	1.90	*0.66*	0.89	17.22	13.42
B68	17.48	17.82	13.15	1.55	16.98	1.87	1.71	1.77	1.85	*0.98*	18.79	4.01
BHM	2. 31	2.05	1.71	13.22	*1.05*	1.22	1.06	1.20	1.38	1.39	2.07	2.02
DHT	1.67	0.69	5.83	13.54	4.25	14.56	14.71	6.62	6.56	6.64	*0.65*	0.61
FHM	15.68	14.69	14.47	*0.87*	16.57	1.08	1.74	1.65	0.93	0.77	14.84	3.15
HFT	1.39	1.76	*1.58*	14.51	1.14	1.06	1.21	1.19	1.02	1.11	14.64	1.12
R18	*0.74*	0.63	0.75	0.81	0.80	0.92	0.88	0.85	0.93	0.85	1.02	3.75
TES	1.74	*0.64*	6.65	14.28	4.38	14.56	17.01	6.68	6.68	6.67	0.75	0.69

The values are expressed in Å. The italic indicate the re-docking case. The rows indicate the ligands, and the columns correspond to the PDB structures

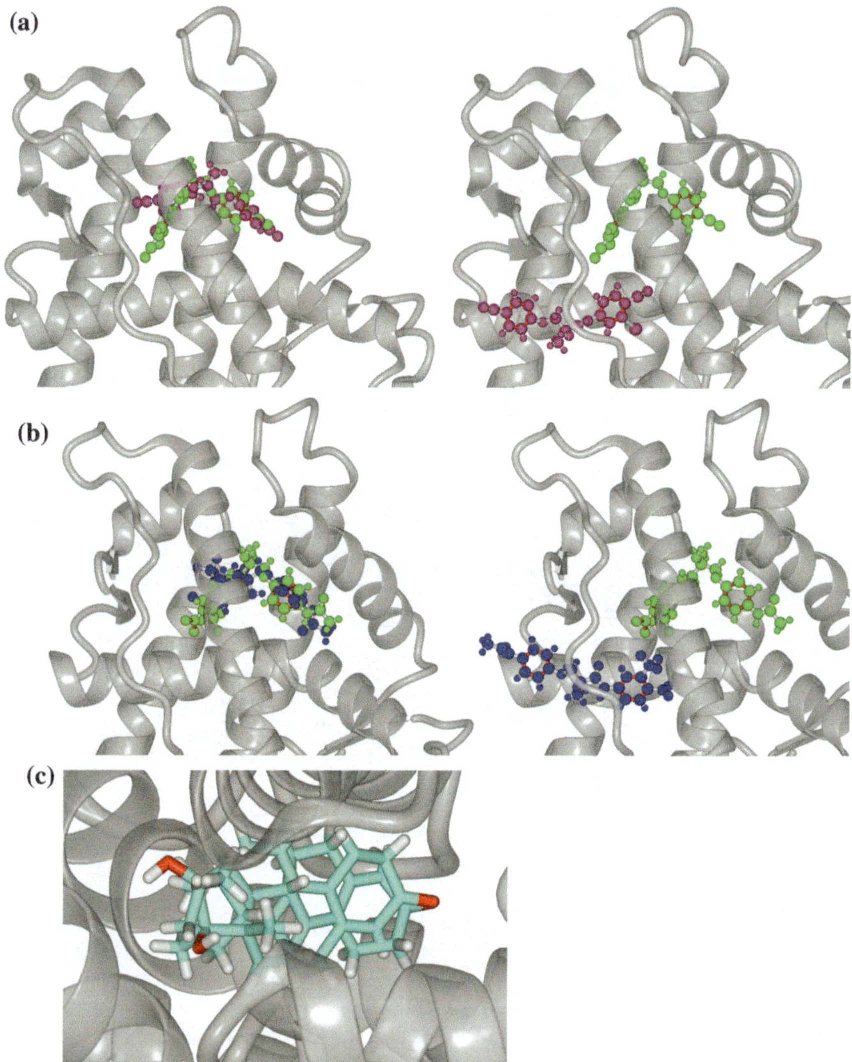

Fig. 2 **a** The best pose HYBR/3B6 superposed with the crystallographic one (3B65) (*left*). The best pose 3L3X/3B6 superposed with the crystallographic one (3B65) (*right*). In *green* is the experimental ligand pose; in magenta is the ligand docked pose. **b** The best pose HYBR/B68 superposed with the crystallographic pose (3B68) (*left*). The best pose 3L3X/B68 superposed with the crystallographic one (3B68) (*right*). In *green* is the experimental ligand pose; in *blue* is the ligand docked pose. **c** The best pose HYBR/R18 superposed with the crystallographic pose (1XOW). The hydroxylated group on the left shows the upside-down conformation of the docked ligand

4 Conclusion

When the crystallographic structure of the complex protein/ligand is available, the position of the binding site is known and the docking procedure is facilitated. In the present work, we compared 11 different AR/ligand complex and the binding pocket was the same in all structures. The binding pocket is also extremely conserved, so the software YADA is particularly adapt to dock an AR agonist. Unfortunately, AR is a particularly flexible receptor and the binding site deforms to adapt to a large variety of ligands. Therefore, it is critical the choice of the receptor geometry, and poor results can results from inaccurate receptor choice. Often, the changes in the binding pocket geometry are overlooked in docking works. Here we presented new protocols to deal with backbone flexibility.

We compared the deformations produced by molecular dynamics, with the ones obtained using of morphing procedure. Molecular dynamics downside is evident for proteins with a very flexible binding site. When unrestrained dynamics is used, the backbone flexibility is too high to prevent the irreversible collapse of the binding site. Therefore, the only available choice is morphing that is capable to deform the binding site even when it is buried inside a protein. The presence of highly mobile and conserved residues can be exploited in restrained MD simulations and it will be the subject of further investigations.

References

1. Zhao, H., Dong, J., Lafleur, K., Nevado, C., Caflisch, A.: Discovery of a novel chemotype of tyrosine kinase inhibitors by fragment-based docking and molecular dynamics. ACS Med. Chem. Lett. 3(10), 834–838 (2012)
2. de Ruyck, J., Brysbaert, G., Blossey, R., Lensink, M.F.: Molecular docking as a popular tool in drug design, an in silico travel. Adv. Appl. Bioinform. Chem. AABC 9, (1–11) (2016)
3. Meng, X.-Y., Zhang, H.-X., Mezei, M., Cui, M.: Molecular docking: a powerful approach for structure-based drug discovery. Curr. Comput. Aided Drug Des. 7(2), 146–157 (2011)
4. Piotto, S., Di Biasi, L., Fino, R., Parisi, R., Sessa, L., Concilio, S.: Yada: a novel tool for molecular docking calculations. J. Comput. Aided Mol. Des. 30(9), 753–759 (2016)
5. Lopez, D.H., Fiol-deRoque, M.A., Noguera-Salvà, M.A., Terés, S., Campana, F., Piotto, S., Castro, J.A., Mohaibes, R.J., Escribá, P.V., Busquets, X.: 2-Hydroxy arachidonic acid: a new non-steroidal anti-inflammatory drug. PLoS ONE 8(8) (2013)
6. Piotto, S., Concilio, S., Bianchino, E., Iannelli, P., López, D.J., Terés, S., Ibarguren, M., Barceló-Coblijn, G., Martin, M.L., Guardiola-Serrano, F., Alonso-Sande, M., Funari, S.S., Busquets, X., Escribá, P.V.: Differential effect of 2-hydroxyoleic acid enantiomers on protein (sphingomyelin synthase) and lipid (membrane) targets. Biochim. et Biophys. Acta—Biomembr. 1838(6), 1628–1637 (2014)
7. Piotto, S., Trapani, A., Bianchino, E., Ibarguren, M., López, D.J., Busquets, X., Concilio, S.: The effect of hydroxylated fatty acid-containing phospholipids in the remodeling of lipid membranes. Biochim. et Biophys. Acta—Biomembr. 1838(6), 1509–1517 (2014)
8. Scrima, M., Di Marino, S., Grimaldi, M., Campana, F., Vitiello, G., Piotto, S.P., D'Errico, G., D'Ursi, A.M.: Structural features of the C8 antiviral peptide in a membrane-mimicking environment. Biochim. et Biophys. Acta—Biomembr. 1838(3), 1010–1018 (2014)

9. Zhao, H., Caflisch, A.: Molecular dynamics in drug design. European journal of medicinal chemistry **91**(4–14 (2015)
10. Di Biasi, L., Fino, R., Parisi, R., Sessa, L., Cattaneo, G., De Santis, A., Iannelli, P., Piotto, S.: Novel algorithm for efficient distribution of molecular docking calculations. Commun. Comput. Inf. Sci. **587**, 65–74 (2016)
11. Tan, M.H.E., Li, J., Xu, H.E., Melcher, K.: Yong, E.-l.: Androgen receptor: structure, role in prostate cancer and drug discovery. Acta Pharmacol. Sin. **36**(1), 3–23 (2015)
12. Ferraldeschi, R., Welti, J., Luo, J., Attard, G., de Bono, J.S.: Targeting the androgen receptor pathway in castration-resistant prostate cancer: progresses and prospects. Oncogene **34**(14), 1745–1757 (2015)
13. Krieger, E., Vriend, G.: YASARA View—molecular graphics for all devices—from smartphones to workstations. Bioinformatics **30**(20), 2981–2982 (2014)
14. Case, D.A., Babin, V., Berryman, J., Betz, R., Cai, Q., Cerutti, D., Cheatham Iii, T., Darden, T., Duke, R., Gohlke, H.: Amber 14 (2014)
15. Piotto, S., Di Biasi, L., Fino, R., Parisi, R., Sessa, L.: Yada: a novel tool for molecular docking calculations. J. Comput. Aided Mol. Des. **30**(9), 753–759 (2016)
16. Piotto, S., Biasi, L.D., Concilio, S., Castiglione, A., Cattaneo, G.: GRIMD: distributed computing for chemists and biologists. Bioinformation **10**(1), 43–47 (2014)
17. Johnson, M., Zaretskaya, I., Raytselis, Y., Merezhuk, Y., McGinnis, S., Madden, T.L.: NCBI BLAST: a better web interface. Nucleic Acids Res. **36**(Web Server issue), W5–9 (2008)
18. UniProt, C.: UniProt: a hub for protein information. Nucleic Acids Res. **43**(Database issue), D204–212 (2015)
19. McWilliam, H., Li, W., Uludag, M., Squizzato, S., Park, Y.M., Buso, N., Cowley, A.P., Lopez, R.: Analysis Tool Web Services from the EMBL-EBI. Nucleic Acids Res. **41**(Web Server issue), W597–600 (2013)
20. Morris, G.M., Huey, R., Lindstrom, W., Sanner, M.F., Belew, R.K., Goodsell, D.S., Olson, A. J.: AutoDock4 and AutoDockTools4: automated docking with selective receptor flexibility. J. Comput. Chem. **30**(16), 2785–2791 (2009)
21. Trott, O., Olson, A.J.: Software news and update AutoDock Vina: improving the speed and accuracy of docking with a new scoring function, efficient optimization, and multithreading. J. Comput. Chem. **31**(2), 455–461 (2010)
22. Friesner, R.A., Banks, J.L., Murphy, R.B., Halgren, T.A., Klicic, J.J., Mainz, D.T., Repasky, M.P., Knoll, E.H., Shelley, M., Perry, J.K., Shaw, D.E., Francis, P., Shenkin, P.S.: Glide: a new approach for rapid, accurate docking and scoring. 1. Method and assessment of docking accuracy. J. Med. Chem. **47**(7), 1739–1749 (2004)

Modelling Giant Lipid Vesicles Designed for Light Energy Transduction

Emiliano Altamura, Francesco Milano, Massimo Trotta, Pasquale Stano and Fabio Mavelli

Abstract In this paper a deterministic kinetic model describing giant lipid vesicles designed for the transduction of light into chemical energy will be presented and discussed. Although the model is based on a simplified mechanism, kinetic constants taken from experimental measurements have been used. The obtained results have shown that giant vesicles encapsulating the Reaction Center in the lipid membrane can exhibit, in suitable experimental conditions, a sensible increase of the internal pH in half an hour under a constant light irradiation.

Keywords Reaction center · Giant vesicles · Light transduction · Kinetic model · pH gradient

1 Introduction

The synthesis of living cells, starting from scratch is one of the most challenging goals in chemistry and biology [1, 2]. As first, this theme was conceived in the field of the origin-of-life on Earth studies [3, 4], but, in the recent years, it has been enforced due to rapid expansion of synthetic biology that has given additional conceptual stimuli and technical tools to this topic [5, 6]. Despite the recent progresses in the implementation of compartmentalized enzymatic systems [7], the primary generation of chemical energy by molecular machineries remains a missing key function. In this paper, we try to bridge this lack, by studying the feasibility of the implementation of protocells, i.e. simplified minimal cells, capable of trans-

E. Altamura · F. Mavelli (✉)
Dipartimento Di Chimica, Università "Aldo Moro" Bari-Italy, Via Orabona 4,
70126 Bari, Italy
e-mail: fabio.mavelli@uniba.it

F. Milano · M. Trotta
CNR-IPCF, Istituto Per I Processi Chimico Fisici, Via Orabona 4, 70126 Bari, Italy

P. Stano
Science Department, Roma Tre University, Viale G. Marconi 446, 00146 Rome, Italy

© Springer International Publishing AG 2018
S. Piotto et al. (eds.), *Advances in Bionanomaterials*,
Lecture Notes in Bioengineering, DOI 10.1007/978-3-319-62027-5_9

ducing light into chemical energy based on the reconstitution of the photosynthetic reaction center (RC) in the membrane of giant unilamellar vesicles (GUVs).

The photosynthetic reaction center (RC) is a transmembrane pigment-protein complex that plays a major role in the photochemical conversion of light into chemical energy in plants, algae and photosynthetic bacteria [8, 9]. It couples light-induced electron transfer to the generation of a proton concentration gradient across a lipid membrane, via reactions involving a quinone molecule that binds two electrons and two protons at its active site. The so-obtained electrochemical gradient can be harnessed to synthesize ATP [9]. It has been previously show [10–12] that the reconstitution of functional, but randomly oriented, RC is possible in conventional (diameter 50–100 nm see Fig. 1) lipid vesicles typically obtained by the detergent depletion method [13].

On the other hand, in recent papers, we have prepared giant lipid vesicle (1–100 μm) by using the droplet transfer method [14, 15] in order to encapsulate in the internal core volume enzymes for cascade reactions [16], suspension of purified biomolecules for the synthesis of proteins and nucleic acids [17]. Furthermore, we have also explored the possibility to decorate the lipid bilayer with self-assembled pores made of hydrophobic molecules in order to improve transportation of substrates across the vesicle membrane [18].

Remarkably, the facile assembly of GUVs, as obtained by the droplet transfer method, paves the way to couple this technique with the procedure for the reconstitution of RC in nano-sized lipid membrane. This can allow the construction of novel and more functional protocells for synthetic biology (RC@GUVs) able to transduce light energy in chemical energy in form pH gradient that ultimately can be exploited for the synthesis of ATP molecules.

This work is part of a series of ongoing experimental [16–18] and theoretical investigations [19–22] on the design and construction of self-sustaining protocells

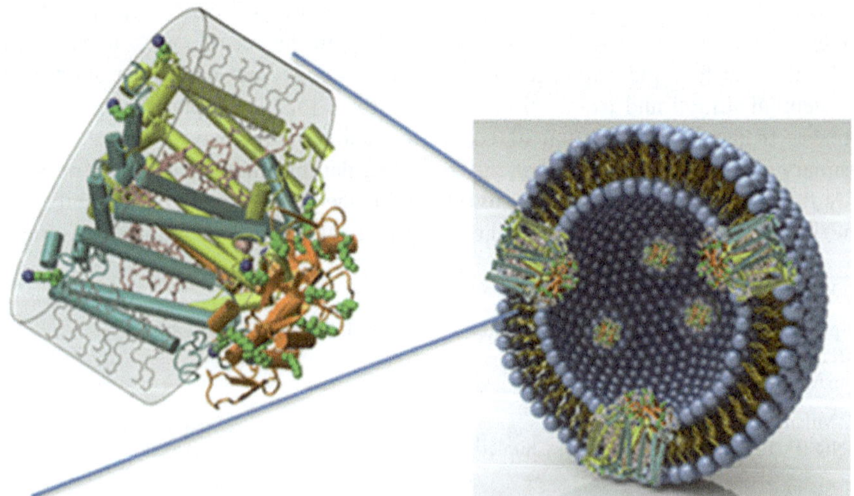

Fig. 1 Physiologically oriented photosynthetic RCs within a nano-sized vesicle membrane

[2]. In particular, here we try to assess theoretically the feasibility of RC@GUVs as transducers of light into chemical energy in form of a pH gradient taking place across the lipid membrane. For this purpose a simplified deterministic model for the reaction center photo-activity will be presented in the next section. This model will be used to predict the rate of the pH increase in the internal aqueous lumen of GUVs under defined operative conditions.

2 The Photo-Conversion Model

In this section, a simplified model for the cyclic photo-conversion performed by RCs will be presented and discussed in details (see Fig. 2). The proposed reaction scheme has been derived from a more detailed stochastic model recently presented by Geyer et al. [23] and further improved from some of the authors [24]. The Geyer's model is focused on the description of the behavior of in vivo chromatophores under different operative conditions and it takes into account also the presence of the enzymatic complexes: *ubiquinol-cytochrome c oxidoreductase* and *F0–F1 ATP sythase* in the membrane of the compartments. On the other hand, the aim of our simplified model is only to make a forecast for the pH time increase in the RC@GUVs internal lumen under fixed experimental conditions: a continuous illumination and the presence of an excess of quinone and cytochrome c^{2+}. In particular we are interested in estimating the time scale of the proton consumption by ranging the size of the GUVs and the RC surface concentration.

2.1 *General Assumptions*

The main assumptions that underlie our theoretical approach are the following: (1) the stochastic effects due to fluctuations in the occurring time of reactive events

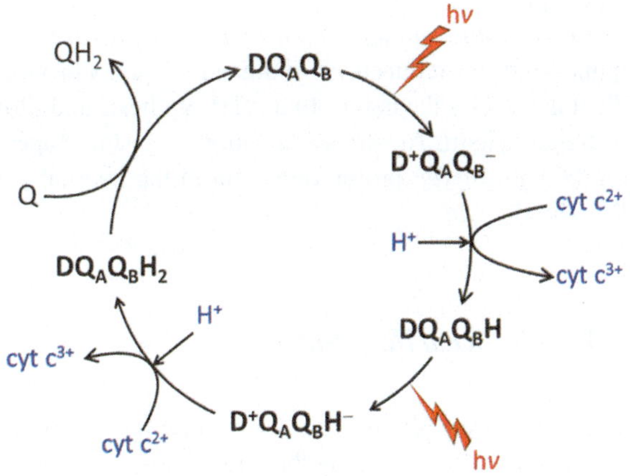

Fig. 2 The RC photocycle. Schematic draw of the reaction network for transducing light into chemical energy producing an increase in the internal pH of GUVs. *Blue* labels evidence aqueous species

can be neglected [25], at least in first approximation, since giant vesicle are very large (with diameter ranging from 10 to 60 µm) compared to natural cromatophores (with diameters ranging from 40 to 60 nm); (2) giant vesicles are considered spherical systems, with poly-dispersed diameters, composed by a lipid bi-layered membrane and the aqueous internal core; (3) diffusion is assumed very fast compare to chemical reactions and it will be neglected in first approximation [23]; (4) all the reactions of the photo-cycle take place in the lipid membrane, where all the reacting specie are embedded except for the oxidized/reduced cytochrome and protons that are bulk aqueous species taken from external and the internal water pool respectively; (5) the free quinone Q is present in excess in the lipid membrane, so that the Q_B site of the reaction center is supposed completely occupied.

2.2 The Photo-Cycle Reaction Network

A sketch of the proposed reaction network for the photo-cycle is reported in Fig. 2. The kinetic mechanism is composed by 5 different main stages that condense many elementary steps in order to simplify the mathematical treatment. Under the assumption that an excess of free quinone is present in the lipid membrane, the process starts when RC (DQ_AQ_B) absorbs a photon and generates an electron-hole couple ($D^+Q_AQ_B^-$). This event triggers the first stage of the photo-cycle: the cytochrome $cyt\ c^{2+}$ present in the external solution donates an electron by reducing the RC, while a proton taken from the internal aqueous solution neutralizes the negative charge on the Q_B site (DQ_AQ_BH). These two stages are repeated when a second photon is absorbed generating again an electron-hole couple ($D^+Q_AQ_B^-H$). This electron-hole intermediate undergoes the same fate as before: it consumes another molecule of $cyt\ c^{2+}$ and binds a second proton, thus producing the quinole in the Q_B site ($DQ_AQ_BH_2$). Therefore, a pH increase is determined in the giant vesicle internal water pool by proton depletion. The quinole is then rapidly replaced by a fresh quinone from the lipid membrane and the photo-cycle can starts again by absorbing a new photon.

In vivo, the *quinol-cytochrome c oxidoreductase* recycles both the quinone/quinol and the reduced c^{2+}/oxidized c^{3+} cytochrome pools, while the proton gradient is used by the cell to fuel ATP synthesis and ultimately the whole metabolism of the organism. As already mentioned, in this paper the attention is focused only on the light-driven proton depletion in the internal vesicle aqueous core due to the RC photo-cycle.

2.3 The Reaction Rates

The mechanism proposed is constituted by two main stages: (a) the electron-hole couple formation driven by the photon absorption and (b) the proton uptake and RC reduction due to the cytochrome c^{2+} oxidation. These two stages are repeated twice in a photo-cycle, reactions (1) and (1′), (2) and (2′) in Fig. 2, in order to form a quinol.

Therefore, a final stage occurs (3) that consists in the replacement of quinol in the $Q_B H_2$ site with a free quinone present in the lipid membrane. This brings back the RC to the initial state of the photo-cycle.

Stages (1) and (1′) can be approximated into elementary steps by the following scheme that takes place in the lipid membrane $\{\}_M$:

$$\{DQ_A Q_B X\}_M \underset{k_{rcg}}{\overset{\sigma_{ph}}{\longleftrightarrow}} \{D^+ Q_A^- Q_B X\}_M \xrightarrow{} k_{eqb} \{D^+ Q_A Q_B^- X\}_M$$

where label X can represent both the precence of a bounded hydrogen or its absence. k_{eqb} is the rate of the electron transfer from Q_A to Q_B while k_{rcg} is the constant for the charge recombination, both constant value are reported in Table 1. σ_{ph} is the rate of charge separation driven by the photon absorption under the experimental conditions. By using the steady state assumption on the intermediate $[D^+ Q_A^- Q_B X]_M$ the rate of the stage can be approximated by the following formula:

$$r_{1/1'} = \frac{k_{eqb}\sigma_{ph}}{k_{rcg} + k_{eqb}} [DQ_A Q_B X]_M \tag{1}$$

$[DQ_A Q_B X]_M$ being the membrane concentration of $DQ_A Q_B$ or $DQ_A Q_B H$ involved in stage (1) and (1′) respectively.

In order to estimate σ_{ph} we consider that the sample can be irradiate through a red filter ($\lambda \geq 600$ nm) using a halogen lamp. The measured irradiance, after the red filter is ca. 200 W/m^2 (DO 9721 Quantum Photo-Radiometer, Delta OHM). It can be shown that, according to the emission spectrum of the source, the irradiance in the 800–900 nm range (where the RC exhibits a absorbance peak) correspond to

Table 1 Rate constants for elementary steps involving the Reaction Center

Reaction	Rate constant	Constant value	References
Charge recombination	k_{rcg}	10 s^{-1}	[12]
Electron transfer from D to Q_b	k_{eqb}	10^{10} s^{-1}	[23]
Binding of reduced cytochrome c_2	k_{cb}	1.4×10^9 nm^3/(s·molecule)	[23]
Reduction of D$^+$ and unbinding of oxidized c_2	k_{cub}	10^3 s^{-1}	[23]
Proton uptake and transfer to Q_b	k_{qh}	10^{10} nm^3/(s·molecule)	[23]
Quinone binding	k_{qb}	1.25×10^5 nm^2/(s·molecule)	[23]
Quinol unbinding	k_{qub}	40 s^{-1}	[23]

ca. 10% of the total irradiance. Thus, taking into account the distance between the lamp and the vesicle suspension, the photon flux density in the 800–900 nm range that reaches each GUV can be estimated by reducing the lamp outgoing flux by an attenuation factor 10^3 obtaining: $0.02/(hc/\lambda_m)$, where h and c are the Plank constant and the speed of light, whereas λ_m is the average wavelength (i.e., 850 nm). This gives a 800–900 nm photon flux density (PFD) of 8.55×10^{16} photons s^{-1} m^{-2}.

As the cross section area of the BChl dimer of about 1.0 nm^2, it can be calculated that each RC is irradiated by about 0.086 photons/s per RC in average. The probability p of photon absorption can be estimated from the RC molar absorption extinction (ε) in the 800–900 region, according to the formula $p = 2.3\ Eq/S$, where E is the molecular extinction coefficient (nm^2) ($E = 1.67 \times 10^{-7} \times \varepsilon$), whereas S (nm^2) is the cross-sectional area of the chromophore (more correctly, the area of density of electron which is responsible for the optical transition), while q is a factor taking into accounts the diverse possible orientation of the chromophore [26]. By using $\varepsilon_{800-900} \approx 1 \times 10^5$ cm^{-1} M^{-1}, $q = 0.3$, $S \approx 1$ nm^2, p is about 1%. This means that under the assumed constant illumination conditions, the rate of charge separation driven by photon absorption is about $\sigma_{ph} = 8.6 \cdot 10^{-4}$ 1/s.

Stages (2) and (2′) represent different elementary steps: (a) the cytochrome binding and unbinding from the external aqueous solution, that determines the reduction $D^+ \rightarrow D$, (b) the proton uptake from the internal water core. These two processes can take place with different order bringing to the formation of DQ_AQ_BXH. Therefore, we assume that the global reaction rate can be calculated by the following formula:

$$r_{2/2'} = \left(\frac{k_{cb}[c^{2+}]_{OUT}}{k_{cub} + k_{cb}[c^{2+}]_{OUT}} \right) k_{qh}[H^+]_{IN} \left[D^+ Q_A Q_B^- X \right]_M \qquad (2)$$

where $[D^+Q_AQ_B^-X]_M$ is now the membrane concentration of the electron-hole couple $D^+Q_AQ_B^-$ or $D^+Q_AQ_B^-H$, while the subscripts OUT and IN indicate aqueous concentrations of the external and the internal water solution respectively. Equation (2) takes into account that in normal conditions the proton uptake is the rate determining step, but the global rate is also modulated by the cytochrome concentration. In fact, if $[c^{2+}]_{OUT} \gg 0$ step (a) does not influence the global rate at all, otherwise when $[c^{2+}]_{OUT}$ tend to 0 step (a) reach a stationary condition.

Finally, the rate of stage (3) is calculated as follows:

$$r_3 = k_{qub}[DQ_AQ_BH_2]_M \qquad (3)$$

assuming that the quinole unbounding is the rate determining steps in the Q_B site repletion, due to the presence of an excess of quinones in the lipid membrane, i.e. assuming that $k_{qb}[Q]_M \gg k_{qub}$.

2.4 The Ordinary Differential Equation Set

The differential equation set (ODES) to be solved in order to get the time evolution of all the reacting species involved in the photo-cycle can be written down as follows:

$$
\begin{aligned}
&(a)\ \frac{d[DQ_AQ_B]_M}{dt} = r_3 - r_1 &\qquad &(g)\ \frac{d[Q]_M}{dt} = -r_3 \\[4pt]
&(b)\ \frac{d[D^+Q_AQ_B^-]_M}{dt} = r_1 - r_2 &\qquad &(h)\ \frac{d[cyt^{2+}]_{OUT}}{dt} = -\frac{S_M}{V_{OUT}}(r_2 + r_{2'}) \\[4pt]
&(c)\ \frac{d[DQ_AQ_BH]_M}{dt} = r_2 - r_{1'} &\qquad &(i)\ \frac{d[cyt^{3+}]_{OUT}}{dt} = \frac{S_M}{V_{OUT}}(r_2 + r_{2'}) \\[4pt]
&(d)\ \frac{d[D^+Q_AQ_B^-H]_M}{dt} = r_{1'} - r_{2'} &\qquad &(l)\ \frac{d[H^+]_{IN}}{dt} = -\frac{S_M}{V_{IN}}(r_2 + r_{2'})\Phi([H^+]_{IN}) \\[4pt]
&(e)\ \frac{d[DQ_AQ_BH_2]_M}{dt} = r_{2'} - r_3 &\qquad & \phantom{(l)\ \frac{d[H^+]_{IN}}{dt}} = -\frac{6}{D_{GUV}}(r_2 + r_{2'})\Phi([H^+]_{IN}) \\[4pt]
&(f)\ \frac{d[QH_2]_M}{dt} = r_3
\end{aligned}
\tag{4}
$$

The ODE set is composed by ten equations seven $(a\text{-}g)$ describing the time evolution of membrane species and three $(h\text{-}i)$ the time course of bulk aqueous species. $[cyt^{2+}]_{OUT}$ and $[cyt^{3+}]_{OUT}$ are present in the external environment, while $[H^+]_{IN}$ is depleted from the internal water core of the giant vesicle. Since the reaction rates r_i are all expressed as surface reaction rates, i.e. the unit measure is $[mol\ dm^{-2}t^{-1}]$, they must must be converted from a rate per surface unit into a rate per unit volume $[mol\ dm^{-3}t^{-1}]$ by multiplying them for the surface/volume ratio S_M/V_{IN}. S_M is the surface area of the giant vesicle membrane, while V_{IN} is the volume of the water pool where the bulk aqueous species are present. Assuming spherical vesicles, the ratio S_M/V_{IN} simply reduces to $6/D_{GUV}$ as reported in Eq. 4 (l), D_{GUV} being the giant vesicle diamenter. On the other hand, for the species located in the external environment V_{OUT} is the GUV co-volume obtained by the reciprocal of the number of vesicles per unit volume:

$$
V_{OUT} = \frac{1}{N_A C_{GUV}}
\tag{5}
$$

where C_{GUV} is the GUV molar concentration and N_A the Avogadro's Number.

Moreover, the function $\Phi([H^+]_{IN})$ in Eq. 4(i) takes into account the acid-base equilibria occurring in the vesicle water pool, as discussed in the next session.

2.5 The Differential Equation of Proton Consumption

In order to couple the differential equations describing the proton depletion in the internal water core with the rate of hydro-quinones and quinoles formation in the lipid membrane, the acid-base equilibria taking place in the vesicle water core are assumed so fast to reach rapidly equilibrium:

$$\mathrm{TrH^+} \overset{K_{Tr}}{\rightleftharpoons} \mathrm{H^+} + \mathrm{Tr} \qquad K_{Tr} = \frac{[H^+]_{IN}[Tr]_{IN}}{[TrH]_{IN}}$$

$$\mathrm{H_2O} \overset{K_W}{\rightleftharpoons} \mathrm{H^+} + \mathrm{OH^-} \qquad K_W = [H^+]_{IN}[OH^-]_{IN} \tag{6}$$

The first equilibrium reported in Eq. 7 is due to the Tris buffer while the last one is the water protolysis. Tris buffer is usually present as chloride salt TrH^+Cl^-, at very low concentration in order to adjust the initial pH value to 7.0 inside the GUVs. The following values have been used for the equilibrium constants: $\mathrm{pK_{Tr}} = 8.2$ and $\mathrm{pK_W} = 14$. Therefore, the differential equation describing the proton comsumption in the vesicle aqueous core can be obtained by writing down the hydrogen/charge balance equation:

$$\frac{6}{D_{GUV}}\left([DQ_AQ_BH]_M + [D^+Q_AQ_{\overline{B}}H]_M + 2[DQ_AQ_BH_2]_M + 2[QH_2]_M\right)$$
$$+ [H^+]_{IN} + [TrH^+]_{IN} = [OH^-]_{IN} + [Cl^-]_{IN} \tag{7}$$

that accounts for the accumulation of negatively charged species in the vesicle lumen due to proton binding by the membrane. By using the acide-base equilibrium equations (Eq. 7) along with mass conservation laws of Tris Buffer: $C_{Tr} = [Tr] + [TrH^+] = [Cl^-]$, one has:

$$\frac{6}{D_{GUV}}\left([DQ_AQ_BH]_M + [D^+Q_AQ_{\overline{B}}H]_M + 2[DQ_AQ_BH_2]_M + 2[QH_2]_M\right) =$$
$$- [H^+]_{IN} + \frac{C_{Tr}K_{Tr}}{K_{Tr} + [H^+]_{IN}} + \frac{K_W}{[H^+]_{IN}} \tag{8}$$

By deriving both members of the previous equations and rembering Eq. 4c–f, finally it is obtained the searched relathionship:

$$\frac{d[H^+]_{IN}}{dt} = -\frac{6}{D_{GUV}} \frac{r_2 + r_{2'}}{1 + \dfrac{C_{Tr}K_{Tr}}{\left(K_{Tr} + [H^+]_{IN}\right)^2} + \dfrac{K_W}{[H^+]_{IN}^2}}$$

3   Results

In this section, the outcomes obtained by numerically solving the ODE set reported in Eq. (5) are presented and discussed. In Table (2) the used parameters are reported. By neglecting the membrane thickness (ca 4 nm), the giant vesicle concentration can be estimated by assuming that all lipids present in the system form vesicles with an average diameter $D_{GUV} = 20$ μm:

Table 2 Parameters used for numerically solving the ODE set reported in Eq. (5)

Parameter	Label	Value
Reaction centre concentration	C_{RC}	2.0 µM
Reduced cytochrome concentration	C_{c2+}	50.0 µM
Quinone bulk concentration	C_Q	50.0 µM
Lipid concentration	C_{LIP}	400 µM
Average vesicle diameter	D_{GUV}	20 µm
POPC lipid head area	α_{LIP}	0.72 nm^2
Initial pH in the vesicle aqueous core	pH_0	7.0

$$C_{GUV} = \frac{C_{LIP}\alpha_{LIP}}{2\pi D_{GUV}^2} \qquad (9)$$

where the factor 2 takes into account the bi-layered nature of the membrane, whereas $S_{MEM} = \pi D_{GUV}^2$ is the vesicle spherical surface. The amphiphile head area is set considering a pure POPC GUV suspension: $\alpha_{LIP} = 0.72$ nm^2. The initial surface concentration of $[DQ_AQ_B]_0 = 0.12$ nmol/dm^2 and $[Q]_0 = 5.78$ nmol/dm^2 are obtained multiplying the bulk concentration, reported in Table (2), by the ratio V_{OUT}/S_{MEM}, where $V_{OUT} = 1.45 \cdot 10^{11}$ dm^3 has been calculated with Eq. (6). In the case of RC the obtained value has been also divided by 2, assuming that proteins are randomly oriented and only 50% are aligned correctly to interact with the external cytochrome.

Moreover, it is worthwhile to stress that the obtained surface concentration is about the 30% of the RC average density in the intra-cytoplasmic membranes of photosynthetic bacteria [27].

In Fig. 3, the time course of the reduced/oxidized cytochrome bulk concentration and the quinone/quinole surface concentration are reported in the upper and lower plot respectively. These graphs clearly evidences that, in the plotted time window, quinone and reduced cytochrome can be considered really in excess accordingly to the made assumptions, since their concentration remains almost constant in time.

On the other hand, Fig. 4 shows that after 30 min the surface concentrations of the different reaction center forms present in the lipid membrane reach a quasi-steady state. The concentration values at the steady state are represented by the black dotted horizontal lines and they can be obtained by equating to zero Eq. 4 (a)–(e):

$$\frac{[DQ_AQ_B]_{M,ST}}{C_{RC}} = \frac{[DQ_AQ_BH]_{M,ST}}{C_{RC}} = \frac{k_{qub}}{\frac{k_{eqb}\sigma_{ph}}{k_{rcg}+k_{eqb}}}\frac{[DQ_AQ_BH_2]_{M,ST}}{C_{RC}} = 0.5$$

$$\frac{[D^+Q_AQ_B^-]_{M,ST}}{C_{RC}} = \frac{[D^+Q_AQ_B^-H]_{M,ST}}{C_{RC}} = \frac{k_{qub}}{\left(\frac{k_{cb}C_{c2+}}{k_{cub}+k_{cb}C_{c2+}}\right)k_{qh}10^{-pH_0}}\frac{[DQ_AQ_BH_2]_{M,ST}}{C_{RC}} = 7.3\times10^{-10}$$

$$(10)$$

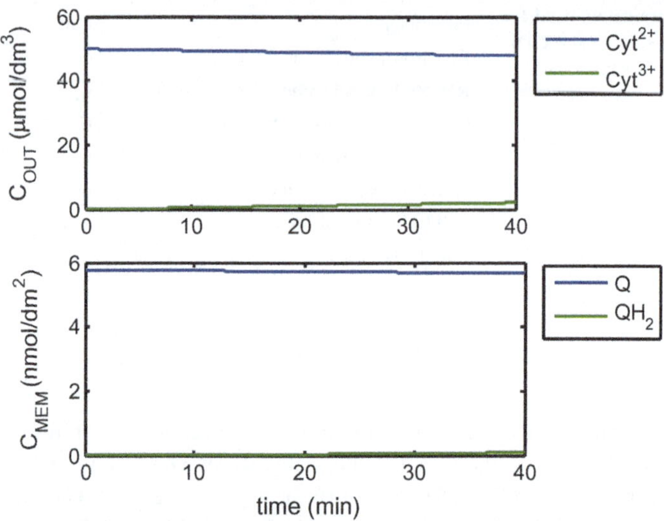

Fig. 3 Time course of the reduced/oxidized cytochrome bulk concentration (*upper plot*) and the quinone/quinole surface concentration (*lower plot*)

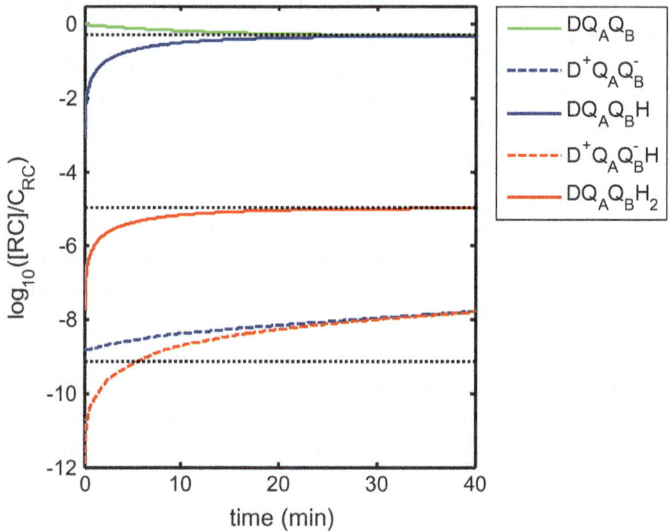

Fig. 4 Time course of different reaction center species in the GUV membrane. The logarithm of surface concentration of each different species divided by the overall concentration C_{RC} is reported against time

and by substituting Eq. (11) in the mass conservation of the RC:

$$\frac{\left[DQ_AQ_BH_2\right]_{M,ST}}{C_{RC}} = \left(1 + 2\frac{k_{qub}}{\frac{k_{eqb}\sigma_{ph}}{k_{rcg}+k_{eqb}}} + 2\frac{k_{qub}}{\left(\frac{k_{cb}C_{c2+}}{k_{cub}+k_{cb}C_{c2+}}\right)k_{qh}10^{-pH_0}}\right)^{-1} = 1.1 \times 10^{-5}$$

$$(11)$$

As it can be seen from the plot in Fig. 4, in the given experimental conditions, the RC is mainly present in the two forms DQ_AQ_B and DQ_AQ_BH (both about 50%), while $DQ_AQ_BH_2$ is present at a much lower fraction: 1.1×10^{-5}. Furthermore, the charged species $^+DQ_AQ_B^-$ and $^+DQ_AQ_B^-H$ are very instable and they are present at a very low fraction. It is import also to remark that Eq. (11) does not predict the steady state value of the charged species satisfactorily, since their concentrations depend on the pH of the internal vesicle solution that continuously change in time.

In Fig. 5, the pH time trend in the internal giant vesicle core is reported for different vesicle diameter. The black line has been calculated using the same parameters as in Figs. 3 and 4, while in order to determine the pH time courses assuming different D_{GUV} also the C_{GUV} and V_{OUT} parameter have been recalculated by using Eqs. (6) and (7). This plot clearly displays that pH increases faster in smaller GUVs. This is due to the fact that, since the surface concentration of correctly aligned RCs remains constant, i.e. the ratio $C_{RC}/(0.5C_{LIP}\alpha_{LIP})$, also the uptake rate of protons per surface unit remains almost constant, while the internal vesicle volume per surface unit increases as $D_{GUV}/6$ and this induces a slower increase in the pH as the vesicle is getting larger.

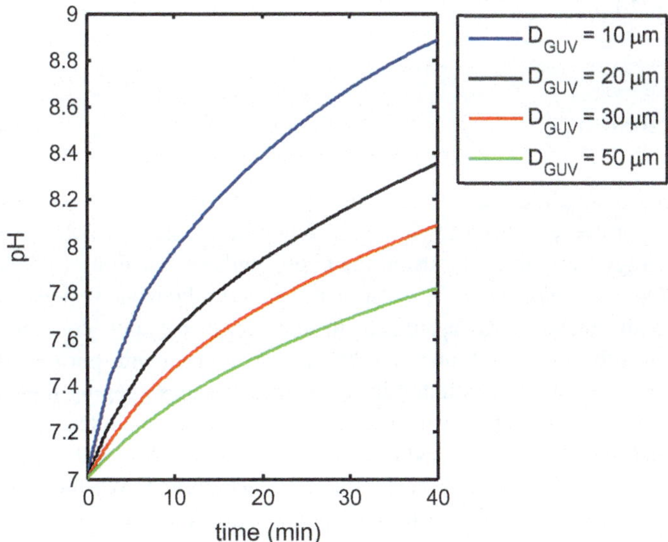

Fig. 5 pH increase in the GUV internal aqueous core depending on the vesicle diameter D_{GUV}

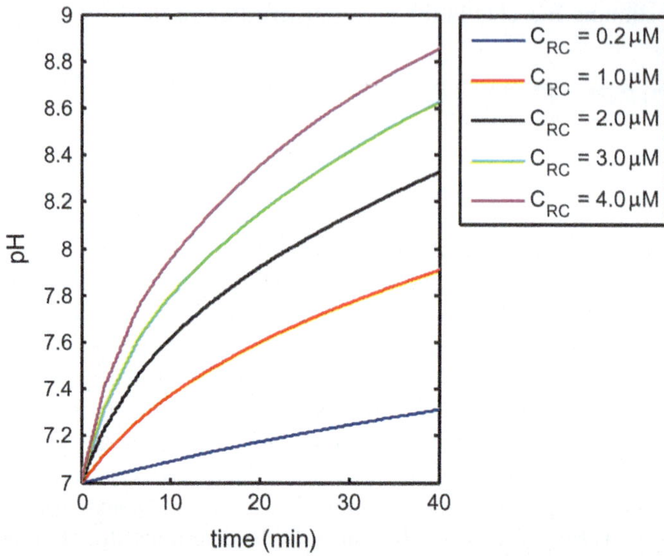

Fig. 6 pH increase in the GUV internal aqueous core depending on the overall RC concentration

In Fig. 6, the time course of the internal pH is plotted for different RC overall concentrations that are linked to the surface concentration by the formula: $[DQ_AQ_B]_0 = 0.5C_{RC}V_{OUT}/S_{MEM}$. As expected, the pH increase is faster when the RC surface is higher. Moreover, even if the surface concentration of active RC is very low (3% of the physiological value) an increase of 0.2 pH units is still observed in 30 min.

4 Conclusion

In this paper, a simplified deterministic model has been presented and discussed in order to assess the feasibility of the implementation of giant vesicles as light into chemical energy transducers by using the photosynthetic reaction centers extracted from *Rhodobacter sphaeroides*. In past years, giant unilamellar vesicles have been prepared with enzymes encapsulated in the internal water core or proteins embedded in the lipid membrane [16–18]. On the other hand, purified RCs have been also successfully reconstituted in nano-sized vesicles keeping their functionality [10, 12]. In this paper, we envisage that RCs reconstituted in giant lipid membrane RC@GUVs can be used as micro sized transducers for the conversion of light energy into a pH gradient across the lipid membrane. We have shown that RC@GUVs can produce, in the internal aqueous vesicle pool, a pH increase of about one unit in less than half an hour when an excess of quinone and reduced cytochrome are added externally. This increment can take place even if the RC surface concentration is lower than the one observed in nature (about 30%) and the sample is continuously irradiated in a not very efficient way (attenuation factor about 10^3). Moreover, smaller giant vesicles have been proved to be much more efficient energy transducers in terms of pH increase thanks to a more favorable volume/surface ratio. We hope that these results will give a strong pulse to the implementation of compartmentalized systems for light transduction by coupling the photosynthetic reaction center and the giant vesicles, thus making a step forward toward the synthesis of autotrophic protocells.

References

1. Noireaux, V., Libchaber, A.: Proc. Natl. Acad. Sci. U.S.A. **101**, 17669–17674 (2004)
2. Luisi, P.L., Ferri, F., Stano, P.: Naturwissenschaften **93**, 1–13 (2006)
3. Morowitz, H.J., Heinz, B., Deamer, D.W.: Orig. Life Evol. Biosph. **18**, 281–287 (1998)
4. Szostak, J. W., Bartel, D. P., Luisi, P. L., Nature, 409, 387–39 (2001)
5. de Lorenzo, V., Danchin, A.: EMBO Rep. **9**, 822–827 (2008)
6. Pohorille, A., Deamer, D.: Trends Biotechnol. **20**, 123–128 (2002)
7. Küchler, A., Yoshimoto, M., Luginbühl, S., Mavelli, F., Walde, P.: Nat. Nanotech. **11**, 409–420 (2016)
8. Allen, J.P., Feher, G., Yeates, T.O., Komiya, H., Rees, D.C.: Proc. Natl. Acad. Sci. U.S.A. **85**, 8487–8491 (1988)
9. Allen, J.P., Williams, J.C.: FEBS Lett. **438**, 5–9 (1988)
10. Nagy, L., Milano, F., Dorogi, M., Trotta, M., Laczkó, G., Szebényi, K., Váró, G., Agostiano, A., Maróti, P.: Biochemistry **43**, 12913–12923 (2004)
11. Milano, F., Italiano, F., Agostiano, A., Trotta, M.: Photosynth. Res. **100**, 107–112 (2009)
12. Mavelli, F., Trotta, M., Ciriaco, F., Agostiano, A., Giotta, L., Italiano, F., Milano, F.: Eur Biophys J Biophys Lett **43**, 301–315 (2014)
13. Ollivon, M., Lesieur, S., Grabielle-Madelmont, C., Paternostre, M.: Biochim. Biophys. Acta **1508**, 34–50 (2000)
14. Stano, P., Carrara, P., Kuruma, Y., Souza, T.P., Luisi, P.L.: J. Mater. Chem. **21**, 18887–18902 (2011)
15. Pautot, S., Frisken, B.J., Weitz, D.A.: Proc. Natl. Acad. Sci. U.S.A. **100**, 10718–10721 (2003)
16. Grotzky, A., Altamura, E., Adamcik, J., Carrara, P., Stano, P., Mavelli, F., Nauser, T., Mezzenga, R., Schluter, A.D., Walde, P.: Langmuir **29**, 10831–10840 (2013)
17. Stano, P., Carrara, P., Kuruma, Y., Souza, T.P., Luisi, P.L.: J. Mater. Chem. **21**, 18887–18902 (2011)
18. Altamura, E., Stano, P., Walde, P., Mavelli, F.: Int. J. Unconv. Comp. **11**, 5–21 (2015)
19. Mavelli, F.: BMC Bioinf **13**(4), S10 (2012)
20. Mavelli, F., Ruiz-Mirazo, K.: Integr Biol **5**, 324–341 (2013)
21. Mavelli, F., Altamura, E., Cassidei, L., Stano, P.: Entropy **16**, 2488–2511 (2014)
22. Mavelli, F., Stano, P.: Artif Life **21**, 1–19 (2015)
23. Geyer, T., Lauck, F., Helms, V.: J. Biotech. **129**, 212–228 (2007)
24. Geyer, T., Lauck, F., Mol, X., Blaß, S., Helms, V.: PLoS ONE **5**, e14070 (2010)
25. Mavelli, F., Piotto, S.: J. Mol. Struct. **771**, 55–64 (2006)
26. Vekshin, N.: Photonics of biopolymers. Springer-Verlag, Berlin Heidelberg (2002)
27. Sener, M.K., Olsen, J.D., Hunter, C.N., Schulten, K.: Proc. Natl. Acad. Sci. U.S.A. **104**, 15723–15728 (2007)

A Computational Approach to the Design of Scaffolds for Bone Tissue Engineering

Antonio Boccaccio, Antonio Emmanuele Uva, Michele Fiorentino, Vitoantonio Bevilacqua, Carmine Pappalettere and Giuseppe Monno

Abstract Design of scaffolds for tissue engineering entails multi-disciplinary and multi-scale aspects. Since in vivo analysis of the tissue regeneration process is quite difficult in terms of selecting experimental protocols and requires considerable amount of time, a variety of numerical models have been developed to simulate mechanisms of tissue differentiation. The tremendous enhancement in computing power led researchers to develop more and more sophisticated models mostly based on finite element techniques and mechano-regulation computational models. In this article, we present an algorithm that combines the finite element model of an open-porous scaffold, a numerical optimization routine and a mechanobiological model. This algorithm has been utilized to determine both, the best scaffold geometry and the best load value (to apply on the scaffold) that allow the bone formation to be maximized.

Keywords Geometry optimization · Scaffold micro-architecture · Bone tissue scaffold · Computational mechanobiology

1 Introduction

Bone tissues undergo continuous remodeling to preserve functionality after injury. Diseases or trauma may cause bone inability to maintain its initial structural properties and even small perturbations may alter bone regeneration processes. In these conditions, surgery is usually required in order to re-establish the structural integrity of the fractured bone.

Bone defects of critical dimensions usually present harsher challenges to successful bone repair. Scaffold implantation hence becomes mandatory as it provides

A. Boccaccio (✉) · A.E. Uva · M. Fiorentino · C. Pappalettere · G. Monno
Dipartimento di Meccanica, Matematica e Management, Politecnico di Bari, Bari, Italy
e-mail: a.boccaccio@poliba.it

V. Bevilacqua
Dipartimento di Ingegneria Elettrica e Dell'Informazione Politecnico di Bari, Bari, Italy

© Springer International Publishing AG 2018
S. Piotto et al. (eds.), *Advances in Bionanomaterials*,
Lecture Notes in Bioengineering, DOI 10.1007/978-3-319-62027-5_10

a porous framework, i.e. an incubator for cell attachment and proliferation. Scaffolds must favorite important biological processes such as vascularization, signaling and production of extra-cellular matrix. Furthermore scaffolds must correctly transfer the mechanical loads to the biological tissues located in the vicinity. Typically, three different groups of biomaterials are utilized in the fabrication of scaffolds for bone tissue engineering: ceramics, synthetic polymers and natural polymers. Each of these biomaterial groups has both, advantages and disadvantages which led to the use of composite scaffolds comprised of different phases [1]. Mismatch of the mechanical properties, between scaffold and surrounding tissue, can lead to stress shielding around the scaffold and subsequently inhibit tissue ingrowth or lead to implant loosening. Therefore, the mechanical properties of scaffolds for bone tissue engineering should be adapted to the mechanical properties of the surrounding tissue.

Complexity of scaffold geometries and large variability of bone tissue properties (e.g. pore size and distribution, mechanical properties, mineral density, cell type and cytokines gradient features), as well as differences in age, nutritional state, activity (mechanical loading) and disease status of individuals, make the design of scaffolds a very challenging task.

The most common approach in bone tissue engineering is the 'trial-and-error' (T&E) approach where the existing design is modified based on the outcome of in vitro or in vivo experiments [2, 3]. Such an approach requires costly protocols and time-consuming experiments [3]. Furthermore, osteogenesis may be very different if it is produced through in vivo or in vitro experiments. For example, a lower porosity stimulates osteogenesis in vitro by suppressing cell proliferation and forcing cell aggregation, whereas a higher porosity and pore size (> 100 μm) results in a greater bone ingrowth for in vivo experiments [4].

Computational models allow to simulate within a certain degree of approximation how the mechanical environment affects tissue differentiation and bone regeneration. Besides the fact that numerical simulations obviously reduces the experimental part of bone tissue engineering, it should be remarked that computational models can be used for carrying out parametric studies that help to fully understand the mechanisms behind tissue differentiation. The accurate knowledge of those mechanisms will help to enhance reliability of scaffolds functionalities [5–7].

The geometry of porous scaffolds has recently been shown to significantly influence the cellular response and the rate of bone tissue regeneration [8–14].

The objective of this study is to optimize the geometry [15] of scaffolds through the combined implementation of finite element models, numerical optimization routines and mechano-regulation algorithms. In detail, an algorithm has been developed to determine both, the best scaffold geometry and the best load value (to apply on the scaffold) that allow the bone formation to be maximized.

2 Materials and Methods

A parametric three-dimensional finite element (FE) model of an open-porous scaffold with circular pores was developed to investigate its structural response under compression loading (Fig. 1). To this purpose, the general purpose finite element software ABAQUS® Version 6.12 (Dassault Systèmes, France) was utilized. Both the scaffold and the granulation tissue were modelled as poroelastic materials. The radius R of the pores has been optimized so as to maximize the

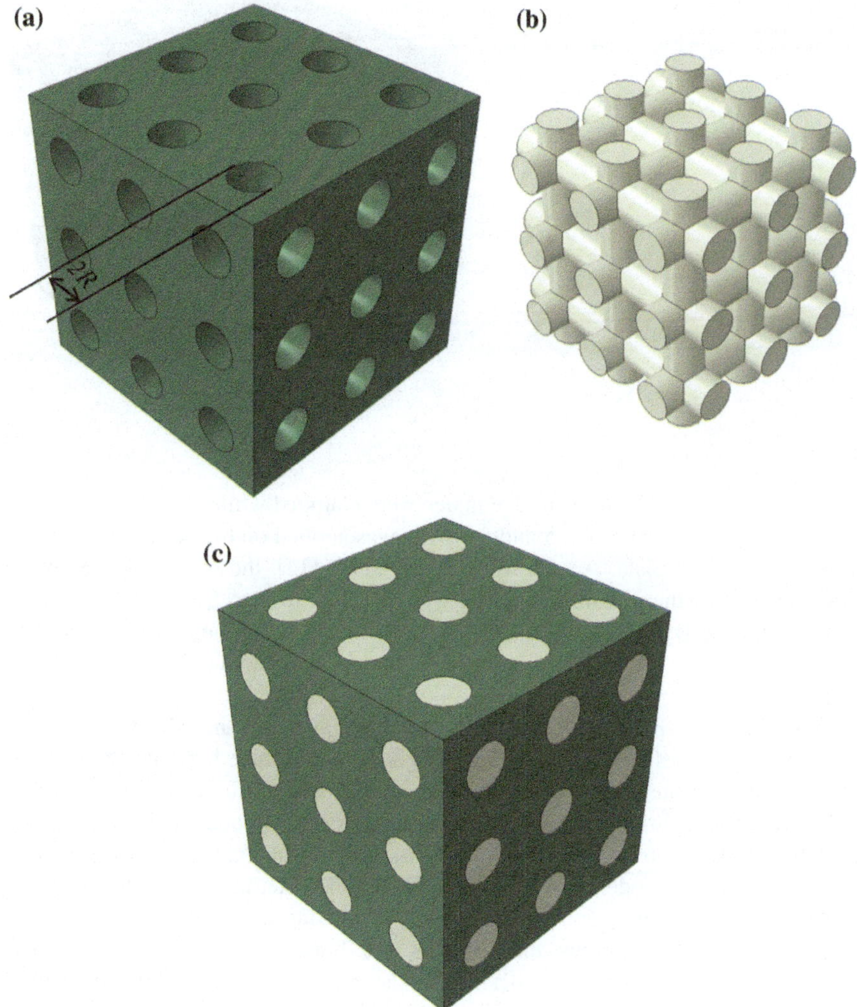

Fig. 1 **a** CAD model of the scaffold (**a**) and of the granulation tissue **b** occupying the scaffold pores. **c** Assembly view of the scaffold system

Table 1 Material properties utilized for the granulation tissue

Material Property	Value
Young's modulus	0.2 MPa
Permeability	1×10^{-14} m^4/N/s
Poisson's ratio	0.167
Porosity	0.8
Bulk modulus grain	2300 MPa
Bulk modulus fluid	2300 MPa

Fig. 2 Finite element mesh of the scaffold-granulation tissue system

amount of bone. The bottom surface nodes were clamped while, a force to generate an apparent pressure of 1.5, 2.5 and 3.5 MPa was applied on the nodes of the upper surface via a rigid plate. According to Byrne et al. [16], the pore pressure on the outer surfaces of the granulation tissue was set equal to zero.

According to Byrne et al. [16], the Young's modulus for the scaffold was fixed equal to 1000 MPa. The material properties used for the granulation tissue are listed in the Table 1.

Mesh included poroelastic elements (C3D4P) available in ABQUS and was properly (Fig. 2) refined so as to reach a good compromise between the computational time and the accuracy of the FEM predictions.

We developed in Matlab environment an algorithm that combines the finite element model above described, a numerical optimization routine and a mechanobiological model (Fig. 3). In each iteration cycle, the algorithm calculates the biophysical stimulus acting on the tissue occupying the scaffold pores.

The biophysical stimulus that triggers the bone regeneration process was assumed to be a function of the octahedral shear strain and of the interstitial fluid flow [17–25]. Following Prendergast et al. [26] if γ is the octahedral shear strain and v the relative fluid/solid velocity, the stimulus S is defined as:

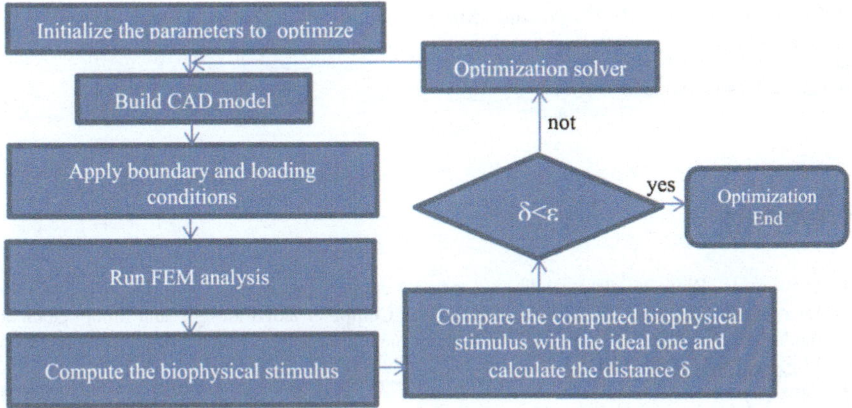

Fig. 3 Schematic of the algorithm utilized to determine the best geometry and the best load value for scaffolds for bone tissue engineering

$$S = \frac{\gamma}{a} + \frac{v}{b} \tag{1}$$

a and b being empirical constants [27], where $a = 3.75\%$ and $b = 3\ \mu ms^{-1}$. The algorithm compares the actual biophysical stimulus obtained in the specific iteration cycle with the ideal one that corresponds to the formation of mature bone and perturbs the scaffold geometry until the difference between these two stimuli is minimized. The optimization problem was solved with the Sequential Quadratic Programming (SQP) method, a globally convergent gradient-based optimization algorithm available in Matlab®. In each cycle of the algorithm, the Size of the Current Search Direction (SCSD) is compared with the Step Size Tolerance (SST). The optimization process terminates when: SCSD < 2SST. In the case the above inequality is not satisfied, the optimization algorithm perturbs scaffold geometry thus commencing a new iteration.

3 Results and Discussion

It was found that the ideal pressure value that can be applied on the scaffold surface with a 200 μm radius circular pores is 0.1777 MPa. In Table 2 the values of R for each of the hypothesized values of the compression loading are listed. Interestingly, the obtained results are consistent with experimental studies reported in the literature [8].

Future research lines should be focused on the development of random-type microstructure geometries as well as on including the angiogenesis process [28] in the optimization algorithm.

Table 2 Values of the radius R predicted by the optimization algorithm for different values of the load

Pressure (MPa)	Radius R (μm)
1.5	190
2.5	175
3.5	150

References

1. O'Brien, F.: Biomaterials & scaffolds for tissue engineering. Mater. today **14**, 88–95 (2011)
2. Lacroix, D., Planell, J.A., Prendergast, P.J.: Computer-aided design and finite element modelling of biomaterial scaffolds for bone tissue engineering. Philos. Trans. A Math. Phys. Eng. Sci. **367**, 1993–2009 (2009)
3. Sanz-Herrera, J.A., García-Aznar, J.M., Doblaré, M.: A mathematical approach to bone tissue engineering. Philos. Trans. A Math. Phys. Eng. Sci. **367**, 2055–2078 (2009)
4. Karageorgiou, V., Kaplan, D.: Porosity of 3D biomaterial scaffolds and osteogenesis. Biomaterials **26**, 5474–5491 (2005)
5. Sun, W., Lal, P.: Recent development on computer aided tissue engineering—a review. Comput. Methods Programs Biomed. **67**, 85–103 (2002)
6. Sun, W., Darling, A., Starly, B., Nam, J.: Computer aided tissue engineering: overview, scope and challenges. Biotechnol. Appl. Biochem. **39**, 29–47 (2004)
7. Sun, W., Starly, B., Darling, A., Gomez, C.: Computer aided tissue engineering: biomimetic modelling and design of tissue engineering. Biotechnol. Appl. Biochem. **39**, 49–58 (2004)
8. Zadpoor, A.A.: Bone tissue regeneration: the role of scaffold geometry. Biomater. Sci. **3**, 231–245 (2015)
9. Rumpler, M., Woesz, A., Dunlop, J.W., van Dongen, J.T., Fratzl, P.: The effect of geometry on three-dimensional tissue growth. J. R. Soc. Interface **5**, 1173–1180 (2008)
10. Bidan, C.M., Kommareddy, K.P., Rumpler, M., Kollmannsberger, P., Bréchet, Y.J., Fratzl, P., Dunlop, J.W.: How linear tension converts to curvature: geometric control of bone tissue growth. PLoS ONE **7**, e36336 (2012)
11. Boccaccio, A., Ballini, A., Pappalettere, C., Tullo, D., Cantore, S., Desiate, A.: Finite element method (FEM), mechanobiology and biomimetic scaffolds in bone tissue engineering. Int. J. Biol. Sci. **7**, 112–132 (2011)
12. Boccaccio, A., Messina, A., Pappalettere, C., Scaraggi, M.: Finite element modelling of bone tissue scaffolds. In: Zhongmin, J. (ed.) Computational Modelling of Biomechanics and Biotribology in the Musculoskeletal System: Biomaterials and Tissues, pp. 485–511 (2014)
13. Guyot, Y., Papantoniou, I., Chai, Y.C., Van Bael, S., Schrooten, J., Geris, L.: A computational model for cell/ECM growth on 3D surfaces using the level set method: a bone tissue engineering case study. Biomech. Model. Mechanobiol. **13**, 1361–1371 (2014)
14. Menolascina, F., Bellomo, D., Maiwald, T., Bevilacqua, V., Ciminelli, C., Paradiso, A., Tommasi, S.: Developing optimal input design strategies in cancer systems biology with applications to microfluidic device engineering. BMC Bioinformatics. **10**(S-12) pp. 4, (2009)
15. Bevilacqua, V., Ivona, F., Cafarchia, D., Marino, F.: An evolutionary optimization method for parameter search in 3d points cloud reconstruction. In: Intelligent Computing Theories. Lecture Notes in Computer Science, vol. 7995 pp. 601–611. Springer (2013)
16. Byrne, D.P., Lacroix, D., Planell, J.A., Kelly, D.J., Prendergast, P.J.: Simulation of tissue differentiation in a scaffold as a function of porosity, Young's modulus and dissolution rate: application of mechanobiological models in tissue engineering. Biomaterials **28**, 5544–5554 (2007)
17. Boccaccio, A., Prendergast, P.J., Pappalettere, C., Kelly, D.J.: Tissue differentiation and bone regeneration in an osteotomized mandible: a computational analysis of the latency period. Med. Biol. Eng. Comput. **46**, 283–298 (2008)

18. Boccaccio, A., Pappalettere, C., Kelly, D.J.: The influence of expansion rates on mandibular distraction osteogenesis: a computational analysis. Ann. Biomed. Eng. **35**, 1940–1960 (2007)
19. Boccaccio, A., Lamberti, L., Pappalettere, C.: Effects of aging on the latency period in mandibular distraction osteogenesis: a computational mechano-biological analysis. J. Mech. Med. Biol. **8**, 203–225 (2008)
20. Boccaccio, A., Kelly, D.J., Pappalettere, C.: A Mechano-Regulation Model of Fracture Repair in Vertebral Bodies. J. Orthop. Res. **29**, 433–443 (2011)
21. Boccaccio, A., Kelly, D.J., Pappalettere, C.: A model of tissue differentiation and bone remodelling in fractured vertebrae treated with minimally invasive percutaneous fixation. Med. Biol. Eng. Comput. **50**, 947–959 (2012)
22. Boccaccio, A., Uva, A.E., Fiorentino, M., Lamberti, L., Monno, G.: A mechanobiology-based algorithm to optimize the microstructure geometry of bone tissue scaffolds. Int. J. Biol. Sci. **12**, 1–17 (2016)
23. Bevilacqua, V., Filograno, G., Fiorentino, M., Uva, A.E. Early diagnosis of lung tumors by genetically optimized 3d-metaball malignancy metric. In Genetic and Evolutionary Computation Conference, GECCO '12, pages 531–538. ACM (2012)
24. Fiorentino, M., Monno, G., Uva, A.E.: Interactive 'touch and see' FEM simulation using augmented reality. Int. J. Eng. Educ. **25**, 1124–1128 (2009)
25. Boccaccio, A., Uva, A.E., Fiorentino, M., Mori, G., Monno, G.: Geometry design optimization of functionally graded scaffolds for bone tissue engineering: a mechanobiological approach. PLoS ONE **11**, e0146935 (2016)
26. Prendergast, P.J., Huiskes, R., Søballe, K.: Biophisical stimuli on cells during tissue differentiation at implant interfaces. J. Biomec. **30**, 539–548 (1997)
27. Huiskes, R., van Driel, W.D., Prendergast, P.J., Søballe, K.: A biomechanical regulatory model of periprosthetic tissue differentiation. J. Mater. Sci. Mater. Med. **8**, 785–788 (1997)
28. Mehdizadeh, H., Sumo, S., Bayrak, E.S., Brey, E.M., Cinar, A.: Three-dimensional modeling of angiogenesis in porous biomaterial scaffolds. Biomaterials **34**, 2875–2887 (2013)

Air Assisted Production of Alginate Beads Using Focusing Flow Microfluidic Devices: Numerical Modeling of Beads Formation

Francesco Marra, Angela De Vivo and Fabrizio Sarghini

Abstract Alginate micro-bead production represents an interesting technological application in many fields such as pharmaceutical, food, and cosmetics. Usually the studies of micro droplet or micro-bead creation in micro channels formed in different geometries and different techniques (mostly T channel or flow-focusing) have been the subject of many research studies using pure, well characterized solutions and do not take into account the behavior and interaction of food grade and natural products. The possibility of using air as focusing flow [2] in microfluidic devices to produce sodium alginate micro-bead introduce some advantages; for example, the utilization of different focusing fluids like oil frequently requires complicate production processes, introducing a barrier to the interaction of alginate solution with the calcium ions during gelification phase and requiring a posteriori filtering and washing procedure. Moreover, direct immersion of liquid alginate drops in a calcium chloride bath to induce gelification usually happens at relatively high speed, inducing a bead shape deformation due to inertial effects. As in microfluidics details really matter, the geometry of the device represents an important issue: small changes in geometrical configuration, like coaxial misalignment could results in major changes in the dynamics of droplet formation. In this work such effects are investigated using numerical analysis.

Keywords Alginate beads · Pre-gelation · Microfluidic · Numerical modeling

F. Marra
Dipartimento di Ingegneria Industriale, Università Degli Studi Di Salerno,
Fisciano, (SA), Italy

A. De Vivo · F. Sarghini (✉)
Dipartimento di Agraria, University of Naples "Federico II", Portici, (NA), Italy
e-mail: fabrizio.sarghini@unina.it

© Springer International Publishing AG 2018
S. Piotto et al. (eds.), *Advances in Bionanomaterials*,
Lecture Notes in Bioengineering, DOI 10.1007/978-3-319-62027-5_11

1 Introduction

Conventional micro and nanoparticles (MP &NP) or encapsulated active compounds fabrication techniques are often characterized by polydispersity and batch-to-batch variations.

The technological framework is plenty of emulsification approaches, mostly involving mixing two liquids in bulk processes, and many of them using turbulence to enhance drop breakup.

In these "top-down" approaches to emulsification, little control over the formation of individual droplets is available and, as a consequence, a broad distribution of sizes is typically obtained, together with a limited loading efficiency control.

This heterogeneity remains a significant obstacle in bulk preparation of drug delivery systems or encapsulation of active compounds for functional foods. Innovations in microfluidics, the science of manipulating fluids in nano/pico-liter scale, introduce several opportunities to improve the fabrication process.

Microfluidic devices can be used both in NP fabrication and for core-shell matrix encapsulation structures. In these processes, the active compounds are driven to the area of adsorption by a carrier enclosed in matrix or core–shell configuration.

The majority of microfluidic methods produce droplet using passive devices generating a uniform, evenly spaced, continuous stream of droplet whose volume ranges from femtoliters to nanoliters.

Their operational modes take advantage of the characteristics of the flow field to deform the interface and promote the natural growth of interfacial instabilities, avoiding in this way the necessity of any local external actuation.

Alginate micro-bead production represents an interesting technological application in many fields such as pharmaceutical, food, and cosmetics in both matrix or mono/poly-nuclear encapsulation.

Industrial scale production of possibly monodispersed alginate beads with desirable characteristics such as controlled size and shape is a desirable target in food industry. Several mechanical and chemical methods had been introduced to produce particles with narrow size distribution in a single step approach, (e.g. milling, oil-water and water-oil emulsions, coacervation, spinning disk, atomization, vibrating nozzle) [6].

Microfluidic devices plays an interesting role in microbeads production due to their relatively cheap technology, and several devices based on different approaches had been proposed; they seem to work quite well on a laboratory scale with fluid mixtures of quite simple rheology and well known properties.

Nonetheless the scale-up of such technologies to the industrial scale often represents a powerful challenge, as several additional characteristics are required: devices should be easy to build, possibly arranged in parallel arrays for mass production with minimal interference between the single devices, the separation between fluids and microbeads should be simple, and possibly cheap.

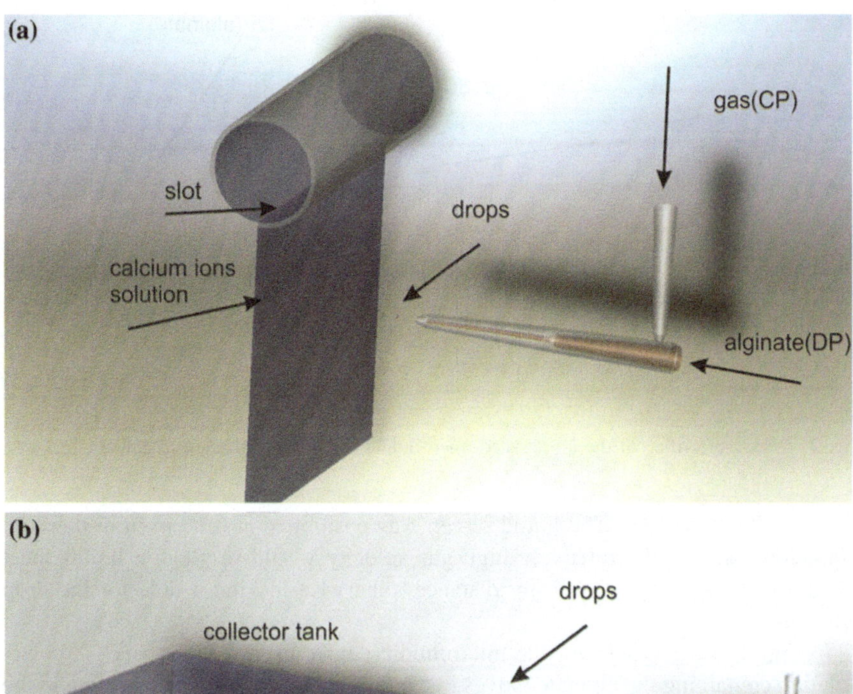

Fig. 1 Sketch of the experimental set-up considered in this work

Among many possibilities, flow focusing is a microfluidic technique where two or more phases of liquid or gases are co-axially focused and then forced through a small orifice. The flow rate of the outer phase, called the continuous phase (CP), usually exceeds that of the inner dispersed phase (DP), typically by ten to thousand times depending on the fluid characteristics. The DP is thus forced into a narrow jet and obliged by DP confinement to flow at the orifice. Due to the rapid change in pressure chamber to the outlet and the prevailing effects of shear stress, fluid dynamics instabilities appear, and the jet breaks up into droplets after passing the orifice [6].

The experimental setup used for this work is described in Fig. 1, where two different configurations (a and b) are depicted: in configuration (a) microbeads produced by the air assisted microfluidic device (shown in Fig. 2) were collected directly by a falling liquid film, acting like a liquid screen obtained using a slot in a polycarbonate horizontal tube; in configuration (b) the micro-beads were recovered after a longer fly using an open collector tank positioned at 1.2 m of distance from

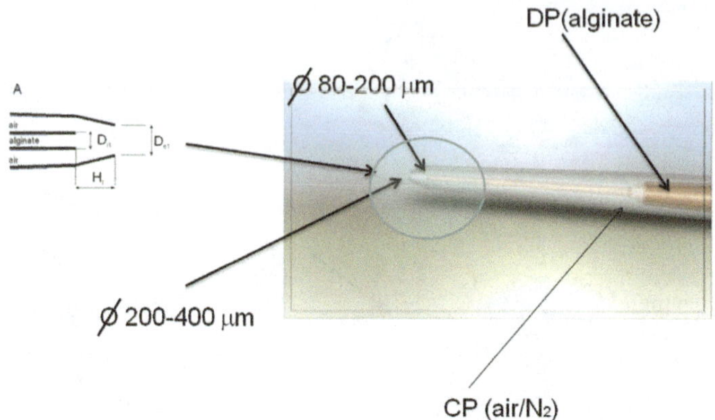

Fig. 2 Details of orifice of the air assisted microfluidic device used in the present study

the microfluidic device. The liquid screen was positioned at 0.15 m from the drop formation zone, as the relatively high gas velocity would result in a liquid sheet break-up if positioned at a nearer distance, otherwise leaving a hole for the drop following.

In Fig. 2, the geometry of the microfluidic device used in this work is shown. Solution containing the alginate moves in the internal tube, while air traveling in the shell is then forced through the orifice (gas-liquid focusing device). In such a device, the main driving force to generate and force the liquid jet out is represented by the gas pressure drop applied to the gas stream, which—in this case—flows with the liquid jet across the orifice of diameter D_e, as described in Fig. 2. The main geometrical parameters are the internal and external diameters (D_i and D_e) respectively of the internal and external orificies, and the capillary-to-orifice distance H.

In order to obtain a better understanding the dynamics of drop formation and a model prevision of their possible shape and dimension at the exit of the microfluidic device, a numerical simulation was performed using computational fluid dynamics.

2 Materials and Methods

Alginate solution (DP) was obtained by purchasing low-viscosity alginic acid sodium salt (obtained from brown algae, Sigma-Aldrich). The aqueous solution used in the experiments was prepared by mixing for 12 h at 30 °C distilled water with a 2% w/v concentration of alginate. The solution was then coloured using UV activated red florescent Createx water based colour for optical inspection. Surface tension for alginate was set to a value of 65 mN/m [1], while rheological properties were obtained experimentally by using a Rheometrics RFS II rheometer. When used in a 2% concentration, the aqueous solution of alginates exhibits a quasi-Newtonian behaviour, while increasing the sodium alginate concentration enhanced the viscosity and shifted the flow behaviour from almost Newtonian to pseudoplastic with thixotropy [9]. In this work a power-law formulation was adopted in numerical simulations, being k = 0.2 the consistency index (Pa s^n) and n = 0.88 is the power law index (dimensionless). Calcium chloride solution for gelation was prepared using AppliChem GmbH Ottoweg Calcium chloride dried powder pure at 97% in a baker stirred on a warm hot plate at 20 °C until the powder was dispersed in distilled water, using concentrations varying from 0.5% w/v to 2% w/v.

Size distribution of obtained alginate micro-beads was determined using a light scattering particle sizer (Mastersizer 3000, Malvern Instruments), and optically with an optical stereomicroscope equipped with a digital camera with 40X magnifying power. Illumination was provided by using a UV lamp at 390 nm wavelength. Mastersizer was used with a laser obscuration greater than 10%; experimental test were repeated three times for each run, and micro-beads were characterized as milky particles with particle adsorption index equal to 1.51 and particle refraction index equal to 1.36. Results were averaged using at least 20 independent measures.

The general equation set governing transport phenomena has been obtained by applying conservation of mass, momentum and energy laws on a control volume in 3D Cartesian coordinates, and it is described in the following lines [8]:

continuity equation

$$\frac{D\overline{\rho}}{Dt} + \overline{\rho}\frac{\partial u_i}{\partial x_i} = 0 \tag{1}$$

conservation of momentum (fluid flow equations)

$$\frac{\partial \overline{\rho}u_j}{\partial t} + \frac{\partial \overline{\rho}\,u_i u_j}{\partial x_i} = \frac{\partial}{\partial x_i}\left[-P\delta_{ij} + \overline{\mu}\left(\frac{\partial u_i}{\partial x_j} + \frac{\partial u_j}{\partial x_i}\right) + \left(\kappa - \frac{2}{3}\overline{\mu}\right)\Delta\delta_{ij}\right] + \overline{\rho}B_j \tag{2}$$
$$j = 1,\ldots 3$$

conservation of energy (heat transfer)

$$c_p\left[\frac{\partial \rho T}{\partial t} + \frac{\partial \rho u_i T}{\partial x_i}\right] = -\frac{\partial q_i}{\partial x_i} + \sigma_{ji}\frac{\partial u_i}{\partial x_j} - P\frac{\partial u_i}{\partial x_i} \tag{3}$$

$$\sigma_{ij} = -P\delta_{ij} + \overline{\mu}\left[\frac{\partial u_i}{\partial x_j} + \frac{\partial u_j}{\partial x_i}\right] + \left(\kappa - \frac{2}{3}\overline{\mu}\right)\Delta\delta_{ij} \tag{4}$$

while for the evolution of the alginate fraction function C (Volume of Fluid)

$$\frac{DC}{Dt} + u_i \frac{\partial C}{\partial x_i} = 0 \tag{5}$$

where, u_i is velocity (m s^{-1}), T is average temperature (K), P is pressure (Pa), ρ is density (kg/m^3), $\mu_{eff} = K\dot{\gamma}^n$ is the effective dynamic viscosity for the alginate solution (Pa s), and c_p is the heat capacity (J kg^{-1} K^{-1}) at constant pressure, and C is the fraction function. All the physical quantities are averaged $(\bar{x} = C\bar{x}_{alg} + (1 - C)\bar{x}_{air})$ depending on the fraction function C of alginate in the local volume. i.e., for each cell, properties such as density ρ and viscosity are calculated by a volume fraction average of all fluids in the cell.

These properties are then used to solve a single momentum equation through the domain, and the numerical velocity field is shared among the fluids.

The numerical solver adopted in this work is OpenFoam (Openfoam v2.2.0, [5]), a CFD software package.

Several mesh configurations were tested, ranging from 2E5 up to 2E6 control volumes.

A Large Eddy Simulation approach was used for turbulence modelling of the gas phase using a scale similar model [7], while CP-DP interaction was analysed by using a Volume of Fluid (VOF) approach [4], as implemented in Denaro and Sarghini [3]. A value of 65 mN/m of relative surface tension between alginate solution and air was set [1], while rheological properties was obtained experimentally by using a Rheometrics RFS II rheometer.

Several ratios between DP and CP were tested: CP mass flow rate ranged from 1 to 1.5 L/min and DP mass flow rate ranged between 0.3 and 0.5 mL/min. Numerical simulations were adopted as a design tool to optimize geometrical configuration of experimental setup and microfluidic device and to analyze jet break-up dynamics providing a droplet size prediction.

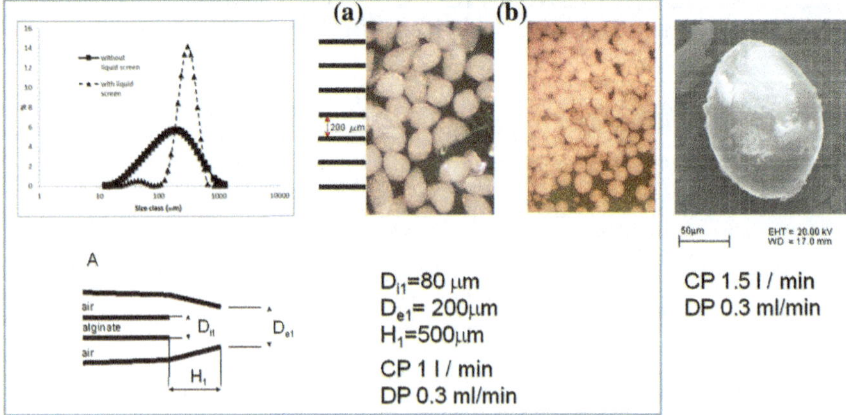

Fig. 3 Distribution of particle size and microbeads obtained, in configuration **a** (jetting on a liquid screen) and configuration **b** (jetting in a collector tank)

3 Results

Before analyzing the results of the numerical analysis, it is opportune to describe the mina results of the experimental activity. In Fig. 3, the distribution of particle size obtained in configuration (a) (jetting on a liquid screen) and configuration (b) (jetting in a collector tank) are shown. The beads obtained in configuration a show a relatively small range of particle size, centered at 350 μm, but bead shape was far from being spherical. Respect to the configuration a, beads collected after the splash into the collector tank without hitting the liquid screen show mean size of 200 μm, with a wider range of particle size (caused by the impact and fragmentation of the drops on the liquid tank surface) but also spherical shape, mainly because of the aerodynamic pressure regularization effects during the flight. The relative high speed at which micro-beads gas-liquid extrusion takes place and the stress related to the beads landing are also responsible of this fragmentation.

In Fig. 4, the dynamic of droplets' formation and flying trajectory is show, at different instants of time (denoted in the figure as t_i). Each image is taken at

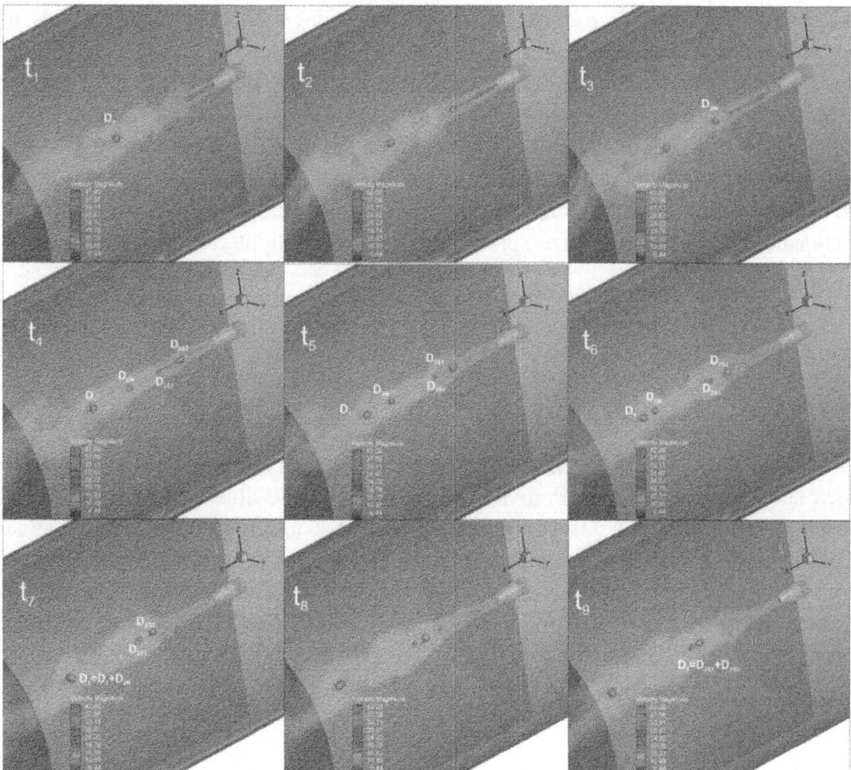

Fig. 4 Dynamic of droplets' formation and flying trajectory

different instant of process time, separated by an interval of 3e-4 s. Since the first instant (t_1) a droplet (D_1) is already present in the investigated domain, while the liquid jet, detached from the tip of the inner capillary, is elongated due to the effect of external disturbance.

Subsequently (time t_2), a droplet D_{2M1} is detached in the frontal part of the jet while the remaining part of the jet undergoes to a secondary elongation and instability, ending with a final formation of two droplets of different sizes, D_{2S1} and D_{2S2}, respectively at times t_3 and t_4. In the domain of interest it is evident the presence of four droplets, each with its own mass and speed: this complex situation brings to further interactions also among droplets themselves. Droplet merging can happen, as in case of D_3, resulting from merging of D_{2M1} with droplet D_1. The reader can follow further evolution in the dynamic of droplets' formation from the pictures in Fig. 4. The final main size of obtained droplets (drop D_3 or D_4 in Fig. 4) is 320 µm, in good agreement with the D_{50} diameter obtained with experiments above described.

The effect of $CaCl_2$ concentration was also investigated, as the availability of Ca^{2+} ions, connected to the concentration of $CaCl_2$ in the aqueous solution which forms the vertical liquid screen, plays an important role in the dynamics of gelation and consequent shape fixation. The dynamics of gelation, which starts from the outer surface of the micro-beads and then move inwards by diffusion, tends to oppose to the successive deformations due to the effects of entrainment of the micro-beads into the liquid screen. These effects can produce either a further breakage of the droplets, resulting in a variation of the dimensional classes' distribution, or in shape regularization due to the interaction of the gelation phase with hydrodynamics effects. Moreover, in 2% w/v $CaCl_2$ concentration, several drops including a tail from incomplete main-satellite separation were also present, increasing the quantitative size spread detected by the light scattering analyser.

4 Conclusions

In this work, the capabilities of numerical analysis applied to modeling of droplets' formation in an air assisted microfluidic device for alginate beads production have been shown and discussed. Particularly, the results were able to describe the proposed configuration (a gas continuous phase plus a liquid sheet to induce an immediate gelation), which represents a viable solution for possible mass production of alginate micro-beads, at least for matrix encapsulation. Quite small and regular micro-beads were produced, although the theoretical lower limit of micro-beads diameter for used device was not investigated in this work.

References

1. Babak, V.G., Skotnikova, E.A., Lukina, I.G., Pelletier, S., Hubert, P., Dellacheriey, E.: Hydrophobically associating alginate derivatives: surface tension properties of their mixed aqueous solutions with oppositely charged surfactants. J. Colloid Interface Sci. **225**, 505–510 (2000)
2. Bong, K.W., Chapin, S.C., Pregibon, D.C., Baah, D., Floyd-Smith, T.M., Doyle, P.S.: Compressed-air flow control system. Lab Chip **11**, 743–747 (2011)
3. Denaro, F.M., Sarghini, F.: 2-d transmitral flows simulation by means of the immersed boundary method on unstructured grids. Int. J. Numer. Meth. Fluids **38**, 1133–1157 (2002)
4. Hirt, C.W., Nichols, B.D.: Volume of fluid (VOF) method for the dynamics of free boundaries. J. Comput. Phys. **39**(1), 201–225 (1981)
5. OpenFoam v.2.2.0, (2013). A free open source CFD code. http://www.openfoam.com/
6. Schneider, T., Chapman, G.H., Häfeli, U.O.: Effects of chemical and physical parameters in the generation of microspheres by Hydrodynamic flow focusing. Colloids Surf., B **87**, 361–368 (2011)
7. Sarghini, F., Piomelli, U., Balaras, E.: Scale-similar models for large eddy simulations. Phys. Fluids **11**(6), 1596–1607 (1999)
8. Sarghini, F.: Microfluidic encapsulation process. In: M. Mishra (ed.) Handbook of Encapsulation and Controlled Release, CRC Press, Boca Raton, USA (2015)
9. Tabeei, A., Samimi, A., Khorram, M., Moghadam, H.: Study pulsating electrospray of non-Newtonian and thixotropic sodium alginate solution. J. Electrost. **70**, 77–82 (2012)

Part III
Applications of Bionanomaterials

Environmental Application of Extra-Framework Oxygen Anions in the Nano-Cages of Mayenite

Adriano Intiso, Raffaele Cucciniello, Stefano Castiglione,
Antonio Proto and Federico Rossi

Abstract Trichloroethylene (TCE) is a chlorinated volatile organic compound (CVOC), used in the last years in dry-cleaning applications and as degreasing agent. In this study we report on the catalytic oxidation of gaseous trichloroethylene (TCE), in a fixed bed reactor, performed by using mayenite ($Ca_{12}Al_{14}O_{33}$) synthetized by using the ceramic method. Results show that mayenite promoted the total oxidation of TCE to carbon dioxide and chlorine in the temperature range 300–500 °C. The catalyst is stable under the investigated reaction conditions showing high recyclability and could be used for several reaction cycles without any loss of activity and selectivity.

Keywords Mayenite · Trichloroethylene · Catalytic oxidation · Chlorinated VOCs

1 Introduction

Trichloroethylene (TCE) is a common chlorinated volatile organic compound and nowadays it has become a dangerous pollutant, being toxic, chemically stable and carcinogenic for humans [1, 2]. TCE is poorly soluble in water, shows high density (>1 g cm^{-3}) and high affinity for organic compounds [3]. TCE is one of the major contaminants of the aquifers [4] and basin on its own physic-chemical properties tends to stratificate at the bottom of groundwater. Actually, several strategies have been developed for the removal of TCE including pump and treat, surface enhanced

Adriano Intiso, Raffaele Cucciniello are contributed equally to this work.

A. Intiso · R. Cucciniello · S. Castiglione · A. Proto · F. Rossi (✉)
Department of Chemistry and Biology, University of Salerno, Via Giovanni Paolo II 132,
84084 Fisciano, SA, Italy
e-mail: frossi@unisa.it

© Springer International Publishing AG 2018
S. Piotto et al. (eds.), *Advances in Bionanomaterials*,
Lecture Notes in Bioengineering, DOI 10.1007/978-3-319-62027-5_12

131

dissolution, adsorption processes, air stripping and phytoremediation [1, 5–7]. Moreover, thermal and catalytic routes were investigated as valuable techniques to TCE complete destruction [8]. Among all the investigated techniques for the elimination of TCE, the catalytic oxidation represents an interesting alternative to the thermal incineration (T > 700 °C) that permits to operate at lower temperatures (250–550 °C) with a drastic reduction of energetic costs.

Furthermore, incineration shows some limitations connected to the chlorinated by-products formation. Recently published works reported the preparation of catalysts able to oxidize and quantitatively convert trichloroethylene, such as metal oxides (MO) and supported noble metals (SMO) [9–11]. MO and SMO show some limitations due to the formation of chlorinated by-products, such as per-chloroethylene (PCE) [12]. Highly active hydrotalcite-like compounds to host metal catalysts, such as Mg(Fe/Al), Ni(Fe/Al) and Co(Fe/Al), have been recently suggested by Blanch-Raga et al. [13]. Catalysts have shown good performances for the TCE oxidation, mainly due to the presence of O_2^- and O_2^{2-} sites that enhance the catalytic productivity, but at the same time, the formation of several by-products (HCl, PCE and Cl_2) were observed. Also many zeolite-like based catalysts were also employed, and they showed a good productivity but a rather low recyclability [14]. In this scenario, many efforts are directed towards the research of new catalysts able to oxidize TCE with high productivity, selectivity and recyclability. The use of noble or transition metals free catalysts has to be preferred to avoid the metals depletion and the high costs for their recovery at the end of the material life-cycle, in agreement with the Green Chemistry principles [15–17].

In our previous works, we have synthesized a mesoporous calcium aluminium oxide, namely mayenite ($Ca_{12}Al_{14}O_{33}$) [18]. The framework of mayenite is composed of interconnected cages with a positively electric charge per unit cell that includes two molecules, $[Ca_{24}Al_{28}O_{64}]^{4+}$, and the remaining two oxide ions O^{2-}, often labelled "free oxygen", were trapped in the cages defined by the framework [19]. Mayenite derivates can be obtained by substituting the free oxygen ions with other anions (Cl^-, H^-, O^-, OH^-, NH_2^-, e^-, etc.) have prepared and used in several research fields [20–22]. In this study we have investigated the ability of O^{2-} ions in the cages of mayenite [23] as active species in the catalytic oxidation of TCE. In the last years, mayenite has already been used in catalysis, for example, Ruszak and coworkers exploited the oxidation ability of mayenite for the selective decomposition of N_2O [24] while Li et al. used mayenite as active support for Ni catalyst for the biomass tar steam reforming [25].

Here we report on the use of pure mayenite as active catalyst for the total oxidation of gaseous TCE. TCE oxidation was carried out in a stainless-steel fixed bed reactor. Mayenite catalyst was prepared by using ceramic method starting from cheap and easily available compounds such as $Ca(OH)_2$ and $Al(OH)_3$. The reaction products were analyzed by means of GC-MS, IC and a IR-based CO_2 probe.

2 Experimental

2.1 *Mayenite Catalyst Preparation and Characterization*

In this study mayenite catalyst was prepared following the ceramic method described by Li et al. [26]. A stoichiometric mixture of $Ca(OH)_2$ and $Al(OH)_3$, (83.00 and 101.4 g respectively), was added to 2 L of distilled water. The mixture was grounded under magnetic stirring (350 rpm) for 4 h at room temperature. The obtained product was separated by filtration and dried at 120 °C overnight, crushed into fine powder and finally placed into a furnace at 1000 °C in air for 4 h.

X-ray powder diffraction was taken with a Bruker D8 Advance automatic diffractometer operating with a nickel-filtered CuKα radiation. Data were collected in the 2θ range of 4–80° with the resolution of 0.02°.

The BET surface area of the catalyst were determined using a Nova Quantachrome 4200e instrument using nitrogen as the probe molecule at liquid nitrogen temperature (−196 °C). Before the adsorption measurement, mayenite sample was degassed at 100 °C under vacuum for 12 h. The surface area values were determined by using 11-point BET analysis.

2.2 *Apparatus and Experimental Conditions*

Catalytic tests were carried out a in a stainless-steel fixed bed reactor (150 mm long, 22 mm o.d, 18 mm i.d) schematically shown in Fig. 1.

The desired mass of the catalyst (0.80 g) was placed into the reactor between two quartz fiber glass layers. The reactor was located in an electrically-heated furnace, and the temperature was measured by a K-thermocouple. Experiments were carried out at different temperature, between 25 and 500 °C. Mayenite was

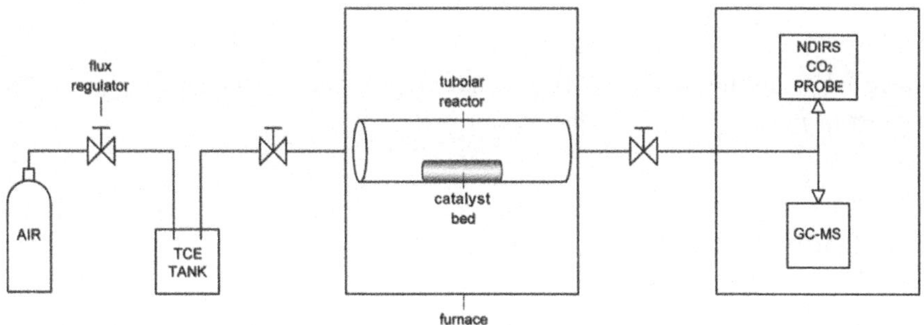

Fig. 1 Experimental setup for trichloroethylene catalytic oxidation

conditioned at the operative temperature with air for 30 min before starting the experiments.

Liquid trichloroethylene (1 μL) was sparged with humid air (RH = 60%) in order to have 1700 ppm of TCE in the gas flow. Catalytic tests have been performed for two different GHSV (Gas Hourly Space Velocity, $GHSV_1 = 60\ h^{-1}$, $GHSV_2 = 12000\ h^{-1}$), in order to evaluate the dependence of the TCE conversion yield by the flow rate.

Blank experiments was also performed in the same reaction conditions but introducing only quartz fiber glass into the reactor without catalyst.

Tedlar sampling bags were used to collect gaseous reaction products before analysis. TCE and other organic compounds of the gas flow were analysed with a GC-MS (Agilent 7890A) equipped with a DB 17-MS column (30 m × 0.25 mm 250 μm). A GC oven was run isothermally at 100 °C for 10 min. Helium was used as carrier gas with a flow rate of 0.5 mL/min and 1.0 mL of sample was analyzed in a split-less injection mode [27].

The concentration of both CO_x and relative humidity was measured by a NDIRS on line probe (Q-Track Plus IAQ Monitor, TSI) placed in the reactor outlet [28]. Cl_2 concentration was determined by SIM mode GC-MS analysis, using the same operative conditions described for VOCs analysis [29]. Furthermore, the presence of Cl_2 was evaluated by means of iodometric titration, by bubbling the effluent stream through a 0.06 M solution of KI and 0.1 M of H_2SO_4 using soluble starch as indicator. The concentration of chloride ions were determined by bubbling the effluent into a 2.6/0.76 mM solution of $NaHCO_3/Na_2CO_3$ and analyzing the resulting chloride amount through ion chromatography (IC) using a Dionex DX 120 (Dionex, Sunnyvale, CA, USA) equipped with a Ion Pac AS14 column (4 mm × 250 mm). Moreover 0.5 g of mayenite were extracted with 3 mL of 2.6/0.76 mM solution of $NaHCO_3/Na_2CO_3$ and analyzed by means of IC with the same experimental conditions used for the gas.

Equation (1) reported the TCE conversion yield, defined as the ratio between the reacted TCE over the total TCE delivered into the reactor:

$$\text{TCE conversion } [\%] = \frac{m^i_{TCE} - m^o_{TCE}}{m^i_{TCE}} \times 100 \tag{1}$$

where m^i_{TCE} are the moles of TCE delivered into the reactor and m^o_{TCE} are the number of moles measured at the reactor outlet.

3 Results and Discussion

Figure 2 presents the XRD patterns of mayenite, prepared according to the ceramic method, by mixing and calcinating a mixture of calcium and aluminium hydroxides.

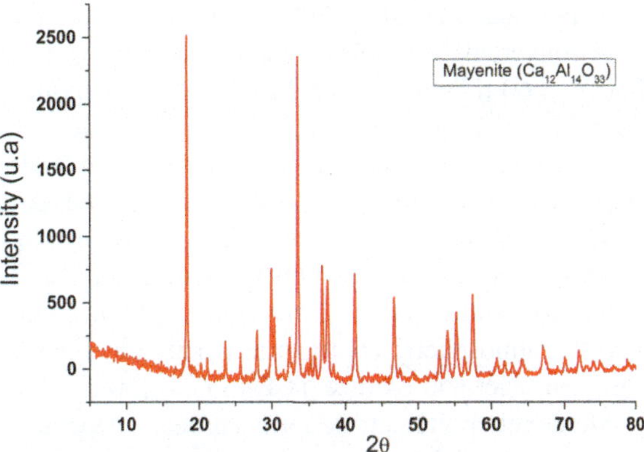

Fig. 2 XRD patterns of calcined mayenite

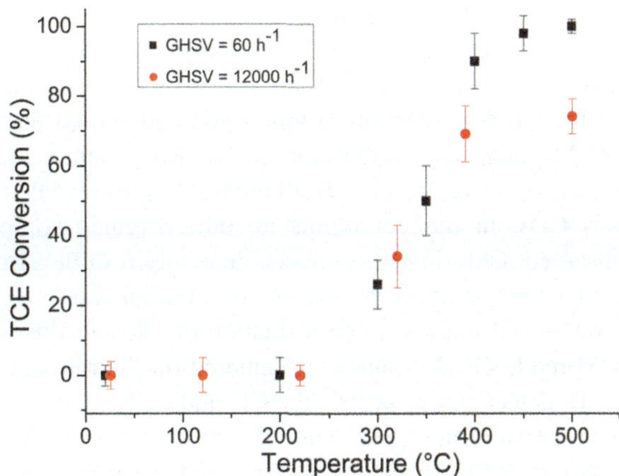

Fig. 3 TCE Conversion yield at different temperatures (*black* square GHSV = 60 h^{-1}, *red* circle GHSV = 12000 h^{-1})

It can be seen the characteristic peaks of mayenite at $2\theta = 18.1°$, $30°$, $33.4°$, $36.7°$, $41.2°$, $46.7°$, $55.2°$ and $57.4°$. $CaAl_2O_4$ and $Ca_3Al_2O_6$ were also detected in trace as common impurities formed during the mayenite synthesis. The BET surface area of the catalyst was 4.5 m^2 g^{-1} in line with those reported in literature for this class of materials.

To evaluate the influence of the GHSV on the TCE conversion yield, we tested the catalytic material by using the reactor reported in Fig. 1 at 60 h^{-1} and 12000 h^{-1}, ranging in the interval 25–500 °C. Furthermore, for all the catalytic tests the concentration of TCE in the flowing air stream was kept constant at 1700 ppm.

Figure 3 shows the TCE degradation yield calculated by means of Eq. 1 for 60 h^{-1} and 12000 h^{-1}.

In each case, for low temperatures (<300 °C) we could not observe a significant degradation of TCE and the concentration of the chlorinated compound at the reactor outlet was found the same as at the reactor inlet. The catalytic activity of the mayenite catalyst starts at 300 °C, where approximately 30% of the delivered TCE was converted. Complete conversion of TCE was reached at 450 °C for GHSV = 60 h^{-1}, where for 12000 h^{-1} the maximum conversion of the pollutant was 75%.

Therefore, as reported in Fig. 3, TCE conversion yield was found to reduce drastically increasing the GHSV. As reported before, the free oxygen species present in the mayenite framework was responsible for the TCE oxidation.

This is in accordance with the temperature and flow rate dependence of the conversion yield; actually, mayenite, at higher temperatures (>400 °C), interacts with the atmospheric O_2 to form reactive oxygen species (O^{2-}, O^-, O_2^-) that, in turn, can react and oxidize TCE to CO_2 [30].

Moreover, pollutant oxidation process depends by the residence time on the catalytic bed of the catalyst. Therefore, at higher GHSV, gaseous TCE have less time to react with the active oxygen species, so decreasing the maximum conversion yield of the pollutant.

Blank experiments, made without using mayenite catalyst at any investigated temperatures, didn't reveal any traces of TCE oxidation.

The organic reaction products were identified and quantified by full spectrum GC-MS analysis. Furthermore, a NDIRS technique was employed to evaluate the presence of CO_x species. Both the techniques revealed that the only carbon-product was CO_2; in our conditions no other organic compound, included CO, wasn't observed. Chlorinated products, generally resulting from TCE oxidation (Cl_2, Cl^-, ClO_x) were analyzed by iodometric titration and ionic chromatography. Iodometric titration did not show the formation of Cl_2 and this sentence was also support by SIM mode GC-MS analysis, centered on the mass m/z = 70.

To detect anion species like Cl^- and ClO_x, both the outlet gases and the catalytic material were analyzed. The effluent stream were bubbled in aqueous solution of bicarbonate/carbonate; at the end of the experiment 20 μL of the solution were injected in an ionic chromatograph. The same analysis was applied for mayenite, after extraction with a carbonate buffer.

IC showed that chloride ions were present only on the catalytic material; no other chlorinated species ClO_x like were detected in the outflow gases or in the catalyst. These results are in accordance with literature where is reported that mayenite tends to adsorb Cl^- to form chloromayenite (Brearleyite, $Ca_{12}Al_{14}O_{32}Cl_2$) due to the replacement of free oxygen species with chlorides [31]. In summary, qualitative analysis of the post-reaction gases and of the catalytic material showed that CO_2 and Cl^- are the mainly products of TCE oxidation, confirmed the data obtained in a previous work [32].

These sentences are in accordance with literature, especially with those reported in a recent review, where the presence of water enhances the reaction selectivity towards HCl and CO_2 [33].

The catalyst stability was tested for ten consecutive experiments in the best operative conditions. At the end of each reaction, mayenite catalyst was restored by

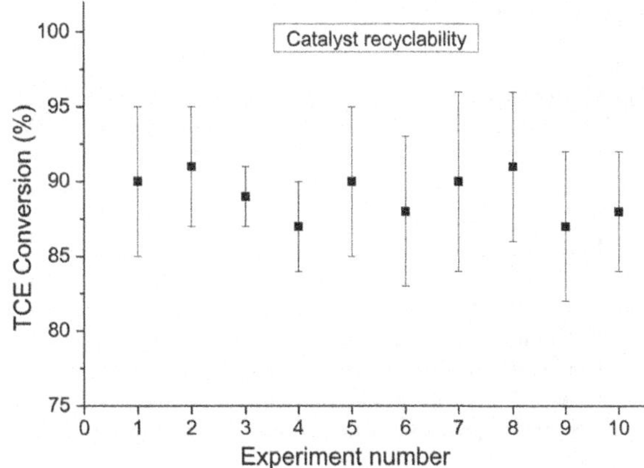

Fig. 4 Recyclability mayenite catalyst (reaction conditions: T_{90} = 400 °C, [TCE]$_{inlet}$ = 1700 ppm, GHSV = 60 h^{-1}, 0.8 g of catalyst)

treating at 450 °C in moist air for 30 min. In this way, the reaction products (e.g. chlorine) could be replaced by free oxygen ions. The results obtained at GHSV = 60 h^{-1} and T = 450 °C are shown in Fig. 4. As it can be seen, the catalyst was stable at this temperature and there was not a significant deactivation after for ten consecutive catalytic cycles (10 h), without any significant loss of activity and selectivity.

These results indicate that mayenite is highly active and stable catalysts for the TCE oxidation. Regarding TCE conversion yield, mayenite showed better performances (90% at 400 °C) than metal un-doped zeolites (<10% at 400 °C) but is less active than metal-loaded zeolites (90–95% at 350 °C) for GHSV = 12000 h^{-1}.

Furthermore, respect to other catalytic systems, mayenite showed several peculiar advantages, namely (i) a complete conversion of TCE in non-toxic species like Cl$^-$ and CO_2, (ii) the existence of reactive oxygen species (O^{2-}, O^-, O_2^-) inside the mayenite that made unnecessary the use of noble and heavy metals, thus reducing the costs and the environmental impact at the end of the material life-cycle.

4 Conclusions

In this work, we showed the prominent use of mayenite as active catalyst for the total oxidation of TCE in the gas phase. TCE is favorably converted, in humid air, with high conversion (>90%) in less hazardous compounds such as CO_2 and HCl at temperature of 460 °C using a GHSV of 60 h^{-1} while a moderate drop in conversion (75%) is observed at GHSV of 12000 h^{-1}. Catalyst shows high stability and recyclability being reused several times without losses in activity and selectivity.

Acknowledgements The authors acknowledge the support by the grants ORSA158121 and ORSA167988 funded by the University of Salerno (FARB ex 60%).

References

1. Greene, H.L., Prakash, D.S., Athota, K.V.: Combined sorbent/catalyst media for destruction of halogenated VOCs. App. Cat. B: Environ. **7**(3–4), 213–224 (1996)
2. Caldwell, J., Blair, A., Bull, R.J., Charbotel, B.: Trichloroethylene, tetrachloroethylene and some other chlorinated agents. IARC Monographs on the evaluation of carcinogenic risks to humans, IARC (International Agency for Research on Cancer) **106**, 1–514 (2014)
3. Rossi, F., Cucciniello, R., Intiso, A., Proto, A., Motta, O., Marchettini, N.: Determination of the trichloroethylene diffusion coefficient in water. AIChE J. **61**(10), 3511–3515 (2015)
4. Fan, A.M.: Trichloroethylene: water contamination and health risk assessment. Reviews of Environmental Contamination and Toxicology, pp. 55–92. Springer, (1988)
5. Boulding, J.R.: EPA Environmental Engineering Sourcebook, pp. 87–100. CRC Press, (1996)
6. Huang, L., Yang, Z., Li, B., Hu, J., Zhang, W., Ying, W.C.: Granular activated carbon adsorption process for removing trichloroethylene from groundwater. AIChE J. **57**(2), 542–550 (2011)
7. Moccia, E., Intiso, A., Cicatelli, A., Proto, A., Guarino, F., Iannece, P., Castiglione, S., Rossi, F.: Use of Zea mays L. in phytoremediation of trichloroethylene. Environ. Sci. Poll. Res. **24**, 11053–11060 (2017)
8. Costanza, J., Mulholland, J., Pennell, K.: Effects of Thermal Treatments on the Chemical Reactivity of Trichloroethylene, US EPA, (2007)
9. Romero-Sáez, M., Divakar, D., Aranzabal, A., González-Velasco, J.R., González-Marcos, J.A.: Catalytic oxidation of trichloroethylene over Fe-ZSM-5: influence of the preparation method on the iron species and the catalytic behavior. App. Cat. B: Environ. **180**, 210–218 (2016)
10. Blanch-Raga, N., Palomares, A.E., Martínez-Triguero, J., Valencia, S.: Cu and Co modified beta zeolite catalysts for the trichloroethylene oxidation. App. Cat. B: Environ. **187**, 90–97 (2016)
11. Divakar, D., Romero-Sáez, M., Pereda-Ayo, B., Aranzabal, A., González-Marcos, J.A., González-Velasco, J.R.: Catalytic oxidation of trichloroethylene over Fe-zeolites. Catal. Today **176**, 357–360 (2011)
12. López-Fonseca, R., Gutiérrez-Ortiz, J.I., González-Velasco, J.R.: Catalytic combustion of chlorinated hydrocarbons over H-BETA and PdO/H-BETA zeolite catalysts. App. Cat. A: Gen. **271**, 39–46 (2004)
13. Blanch-Raga, N., Palomares, A.E., Martínez-Triguero, J., Puche, M., Fetter, G., Bosch, P.: The oxidation of trichloroethylene over different mixed oxides derived from hydrotalcites. App. Cat. B: Environ. **160**, 129–134 (2014)
14. Aranzabal, A., Romero-Sáez, M., Elizundia, U., González-Velasco, J.R., González-Marcos, J.A.: Deactivation of H-zeolites during catalytic oxidation of trichloroethylene. J. Catal. **296**, 165–174 (2012)
15. Gawande, M.B., Bonifácio, V.D.B., Luque, R., Branco, P.S., Varma, R.S.: Solvent-free and catalysts-free chemistry: a benign pathway to sustainability. Chem. Sus. Chem. **7**, 24–44 (2014)
16. Cespi, D., Cucciniello, R., Ricciardi, M., Capacchione, C., Vassura, I., Passarini, F., Proto, A.: A simplified early stage assessment of process intensification: glycidol as a value-added product from epichlorohydrin industry wastes. Green Chem. **18**, 4559–4570 (2016)
17. Cucciniello, R., Pironti, C., Capacchione, C., Proto, A., Di Serio, M.: Efficient and selective conversion of glycidol to 1, 2-propanediol over Pd/C catalyst. Catal. Comm. **77**, 98–102 (2016)
18. Cucciniello, R., Proto, A., Alfano, D., Motta, O.: Synthesis, characterization and field evaluation of a new calcium-based CO_2 absorbent for radial diffusive sampler. Atmos. Environ. **60**, 82–87 (2012)
19. Cucciniello, R., Proto, A., Rossi, F., Motta, O.: Mayenite based supports for atmospheric NOx sampling. Atmos. Environ. **79**, 666–671 (2013)

20. Kitano, M., Inoue, Y., Yamazaki, Y., Hayashi, F., Kanbara, S., Matsuishi, S., Yokohama, T., Kim, S., Hara, M., Hosono, H.: Ammonia synthesis using a stable electride as an electron donor and reversible hydrogen store. Nature Chem. **4**, 934–940 (2012)
21. Hayashi, F., Tomota, Y., Kitano, M., Toda, Y., Yokoyama, T., Hosono, H.: NH_2–dianion entrapped in a nanoporous $12CaO \cdot 7Al_2O_3$ crystal by ammonothermal treatment: reaction pathways, dynamics, and chemical stability. J. Am. Chem. Soc. **136**, 11698–11706 (2014)
22. Proto, A., Cucciniello, R., Genga, A., Capacchione, C.: A study on the catalytic hydrogenation of aldehydes using mayenite as active support for palladium. Cat. Comm. **68**, 41–45 (2015)
23. Lacerda, M., Irvine, J.T.S, Glasser, F.P., West, A.R.: High oxide ion conductivity in $Ca_{12}Al_{14}O_{33}$. Nature **332**, 525–526 (1988)
24. Ruszak, M., Inger, M., Witkowski, S., Wilk, M., Kotarba, A., Sojka, Z.: Selective N_2O removal from the process gas of nitric acid plants over ceramic $12CaO \cdot 7Al_2O_3$ catalyst. Catal. Lett. **126**, 72–77 (2008)
25. Li, C., Hirabayashi, D., Suzuki, K.: A crucial role of O_2^- and O_2^{2-} on mayenite structure for biomass tar steam reforming over $Ni/Ca_{12}Al_{14}O_{33}$. App. Cat. B: Environ. **88**, 351–360 (2009)
26. Li, C., Hirabayashi, D., Suzuki, K.: Synthesis of higher surface area mayenite by hydrothermal method. Mat. Res. Bull. **46**, 1307–1310 (2011)
27. Cucciniello, R., Proto, A., Rossi, F., Marchettini, N., Motta, O.: An improved method for BTEX extraction from charcoal. Anal. Methods **7**, 4811–4815 (2015)
28. Proto, A., Cucciniello, R., Rossi, F.: Motta, O: Stable carbon isotope ratio in atmospheric CO_2 collected by new diffusive devices. Environ. Sci. Poll. Res. **21**, 3182–3186 (2013)
29. Mendez, M., Ciuraru, R., Gosselin, S., Batut, S., Visez, N., Petitprez, D.: Reactivity of chlorine radical with submicron palmitic acid particles: kinetic measurements and product identification. Atmos. Chem. Phys. **13**, 11661–11673 (2013)
30. Teusner, M., De Souza, R.A., Krause, H., Ebbinghaus, S.G., Belghoul, B., Martin, M.: Oxygen diffusion in mayenite. J. Phys. Chem. C **119**, 9721–9727 (2015)
31. Schmidt, A., Lerch, M., Eufinger, J.-P., Janek, J., Tranca, I., Islam, M.M., Bredow, T., Dolle, R., Wiemöfer, H.D., Boysen, H., Hölzel, M.: Solid State Ionics **254**, 48–58 (2014)
32. Cucciniello, R., Intiso, A., Castiglione, S., Genga, A., Proto, A., Rossi, F.: Total oxidation of trichloroethylene over mayenite ($Ca_{12}Al_{14}O_{33}$) catalyst. App. Cat. B: Environ. **204**, 167–172 (2017)
33. Aranzabal, A., Pereda-Ayo, B., Pilar Gonzalez Marcos, M., González-Marcos, J.A., López-Fonseca, R., González-Velasco, J.R.: State of the art in catalytic oxidation of chlorinated volatile organic compounds. Chemi. Pap. **68**, 1169–1186 (2014)

Current Directions in Synthetic Cell Research

Pasquale Stano, Giordano Rampioni, Francesca D'Angelo,
Emiliano Altamura, Fabio Mavelli, Roberto Marangoni,
Federico Rossi and Luisa Damiano

Abstract The construction of synthetic cells of minimal complexity is today one of the most attractive and challenging goals in synthetic biology. Synthetic cells are assembled by combining the methods of liposome technology and microfluidics, and components of cell-free systems. In this contribution, we will shortly illustrate the state-of-the-art of synthetic cell research; next we will present some current trends and scenarios, that could lead to a qualitative jump in the next years. In particular, we will focus on the construction of novel multi-compartment vesicles, the achieving of new functions via the reconstitution membrane-bound proteins, the shift from the isolated cell—to the cell population/community perspective, the exploitation of chemical signalling, and the integration of stochastic mathematical models. The ambitious goal of approaching embodied and minimal cognition from this experimental perspective is also shortly mentioned.

keywords Synthetic cell · Synthetic biology · Lipid vesicles · Origins of life

P. Stano (✉) · G. Rampioni · F. D'Angelo
Sciences Department, Roma Tre University, Rome, Italy
e-mail: pasquale.stano@unisalento.it

Present address:
Biological and Environmental Sciences and Technologies Department (DiSTeBA),
University of Salento, Ecotekne, 73100 Lecce, Italy

E. Altamura · F. Mavelli
Chemistry Department, University of Bari, Bari, Italy

R. Marangoni
Biology Department, University of Pisa, Pisa, Italy

F. Rossi
Department of Chemistry and Biology, University of Salerno, Salerno, Italy

L. Damiano
ESARG (Epistemology of the Sciences of the Artificial Research Group),
Department of Ancient and Modern Civilizations, University of Messina, Messina, Italy

L. Damiano
CERCO (Centre for Research on Complex Systems), University of Bergamo, Bergamo, Italy

© Springer International Publishing AG 2018
S. Piotto et al. (eds.), *Advances in Bionanomaterials*,
Lecture Notes in Bioengineering, DOI 10.1007/978-3-319-62027-5_13

1 What Are Synthetic Cells

Synthetic Cells (SCs) can be defined as those cell-like molecular systems constructed in the laboratory by inserting biological or synthetic molecules inside and on the surface of lipid vesicles (liposomes, see Fig. 1). Ideally, SCs mimic biological cells with respect to structure, functions, interactions, and any other possible property (including the ultimate and long-term goal of being alive) with *minimal complexity*.

Historically, SCs have been first explored to shed light on some origin of life questions, but it became clear that they can be created and adapted for investigating biological mechanism or developing biotechnological tools (Fig. 1). SCs are relevant in origin of life studies [26] because they are useful models of primitive cells (protocells), especially when they are built with allegedly primitive membrane forming compounds (e.g., fatty acids), and host compounds and reactive systems of primitive importance (ribozymes, simple peptides, self-replicating molecules, etc.) [52]. One of the goal is to show that simple living cells, maybe imperfect and "limping" ones, can however emerge from the self-organization of non-living molecules.

On the other hand, synthetic cells can be used to reconstruct, in realistic cell-like architecture, those biological processes that can be advantageously studied in a

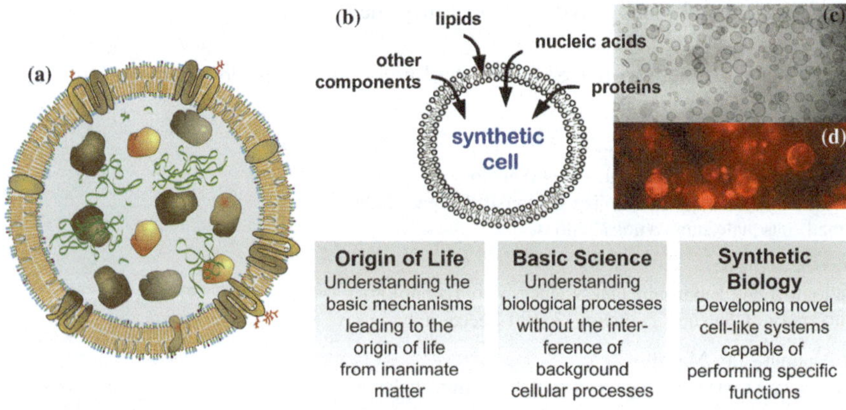

Fig. 1 Schematic drawing of a liposome-based synthetic cell. **a** Synthetic cells are built by incorporating in the aqueous core and in the membrane of lipid vesicles (diameter 0.1–10 μm) the minimal number of molecular components in order to obtain a cell-like structure that performs specific function (ultimately: being alive). The image is taken from [54] with permission from Elsevier [the image has been adapted to represent a synthetic cell, but it actually sketches a biological exosome]. **b** Synthetic cells (or, more correctly, semi-synthetic, *cf.* [26]) can be built by incorporating nucleic acids, proteins, ribosomes and other compounds inside and on the surface of a lipid vesicle. Synthetic cells of conventional size (i.e., diameter below 0.5 μm) can be conveniently imaged by electron microscopy (**c**), whereas *giant* ones (i.e., diameter: 1–50 μm) can be seen by optical microscopy. **d** Synthetic cells can be adapted to perform research in origin of life and basic science, and also be used as biotechnological tools according to synthetic biology

simplified platform. In fact, by reconstructing a certain process in a synthetic cell, its understanding results easier because the interfering "noise" of background cellular processes, as unwanted interactions with other components are purposely removed. The process of interest can be finely tuned owing to the control of synthetic cell composition whereas such control is impossible in living cells.

Finally, SCs can be used as tools for biotechnological applications. Their design and construction is developed under the synthetic biology paradigm, which includes the use of standard parts (http://parts.igem.org), the concepts of modularity and orthogonality [13], and fully programmability (at least in principle). Such bioengineering perspective has led to sketch the possible use of SCs in nanomedicine [25] or in applications of biotechnological flavours, such as sensoring, analysis, programmable on-chip chemical devices, and so on [32, 41].

Starting from the 1990s, several research groups have been involved in synthetic cell research, contributing to setting up the stage for actual and future developments (for reviews, see [47, 49]). It must be noticed that in addition to liposome-based SCs, other artificial compartments have been also explored (*e.g.*, water-in-oil droplets, polymersomes, colloidosomes, coacervates, micelles, oil-in-water droplets).

1.1 From the Pioneer Phase (up to 2001) to the Most Recent Developments

After the 1989 paper on the use of reverse micelles as minimal autopoietic systems [28], early investigation was primarily reported by ETH-Zürich team (Pier Luigi Luisi, Peter Walde and Thomas Oberholzer), which carried out pioneering research on vesicle self-reproduction [55] and performing important proof-of-principle experiments on micro-compartmentalized reactions, namely RNA replication [37], polymerase chain reaction [35], and polypeptide synthesis [36]. Actually, most of the current experimental strategies stems from these seminal papers, with the exception of the recently introduced droplet transfer method and microfluidics-based procedures. The end of the pioneering phase can be symbolically placed in 2001, when the production of a well-folded and thus functional protein inside liposomes was achieved by the Yomo group [58]. Protein synthesis inside liposomes plays a strategic role in SCs research because the proteins carry functions, e.g., they are enzymes, transcriptional factors, membrane channels, and so on.

In the following ten years or so, several fundamental aspects of compartmentalized protein synthesis were explored, and important milestones were reached: (a) the prolongation of intraliposome transcription-translation (TX-TL) reactions for about four days owing to the generation of a α-hemolysin based membrane pore (from inside) [34]; (b) the realization of two-steps genetic cascade reactions inside liposomes [19]; (c) the first report on the synthesis of hydrophobic enzymes (membrane embedded) [22]; (d) the RNA replication via a self-encoded mechanism [21]; (e) the spontaneous formation of super-filled vesicles [27, 48] (Fig. 2); (f) the

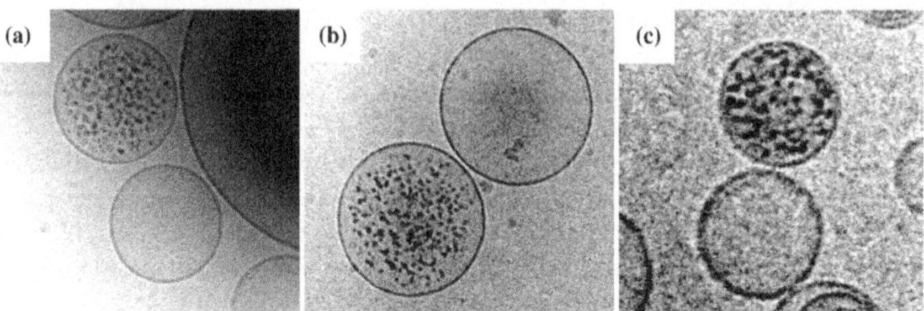

Fig. 2 Cryo-transmission electron microscopy images of ferritin-containing lipid vesicles. Super-filled vesicles coexist with empty vesicles. The internal ferritin concentration reaches high values (in these vesicles about 4–12 times the expected concentration). The diameter of ferritin-filled vesicles are 195, 219, 118 nm, respectively, for picture (**a**), (**b**), and (**c**). A similar behavior is observed with ribosomes [46] and with the full TX-TL machinery (PURE system), see [48]. The fact that solute-filled vesicles form spontaneously might help to explain the origin of solute-rich primitive cellular structures. Reproduced from [27] with permission from Wiley

droplet transfer method [34, 39, 57], which can be used to produce solute-filled giant vesicles with good yields and high encapsulation efficiency; and (g) the ongoing development of microfluidic devices for high-throughput production, in highly reproducible manner, of solute-filled giant vesicles [12]. Many of these results have been obtained by employing the PURE system [45], a standardized and minimal set of macromolecules for synthesizing proteins inside liposomes.

All this—and many other advancements which are omitted here due to space limitation—has brought to a kind of maturation of the field, which now attracts more and more researchers. In other words, times are probably ripe for a sort of qualitative jump.

2 Current Directions in SC Research

What are the current directions in SC research? Here we report a number of topics that we consider important, and that mirror our attitudes, without the presumption of being exhaustive.

2.1 Reconstitution of Advanced Architecture and Functions in SCs

Vesosomes. Whereas current SCs are generally based on a single lipid compartment, generally containing just one lipid membrane (unilamellar lipid vesicle), novel architectures could derive from including small vesicles inside a larger one

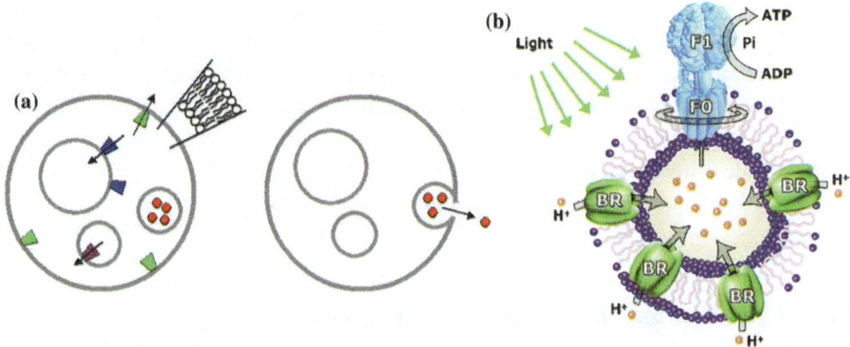

Fig. 3 **a** Vesosomes are multi-vesicular vesicles composed by smaller vesicles inside a larger vesicle. By inserting compartments inside compartments, interesting vectorial properties might arise from oriented membrane proteins. Moreover, the content of internal vesicles can be released in the external medium after membrane fusion. **b** ATP has been produced by coupled reactions between bacteriorhodopsin, a light-driven transmembrane proton pump, and F_0F_1-ATP synthase motor protein, reconstituted in polymersomes. Reprinted with permission from [6]. Copyright 2005 American Chemical Society

(Fig. 3a). Technically, these are called multivesicular vesicles or vesosomes, and it is not rare to find them in vesicle samples formed by classical methods. However, when vesosomes are specifically needed in order to assembly synthetic cells with hierarchical topology (vesicles inside vesicles), their preparation should follow dedicated methods. The construction of vesosomes follows specific protocols [5, 38] but their employment as synthetic cell has not been explored in detail (however, see [16]). The interest in this architecture stems from the cytomimetic role of internalized vesicles: they could play the role of intracellular organelles, allowing novel SC designs in terms of sub-compartmentalization, in particular for exploiting the concentration gradients across their membranes, and expanding the intracellular total membrane area (which will possibly include membrane-bound machineries). Moreover, small vesicles could fuse with the outermost membrane to release their content (a signal molecule for instance) outside the SC in programmable way.

Membrane proteins. Until now, the fundamental relevance of membrane embedded devices (proteins) has been only partially addressed. This is due to the difficulty of reconstituting functional membrane proteins or of synthesizing them directly by SCs. These difficulties are both intrinsic (poor understanding of mechanism of membrane protein folding, refolding, membrane insertion, lipid-membrane interactions) and practical (conflict between required experimental conditions). An early attempt of direct synthesis inside SCs has been reported [22, 44]. Two acyltransferase proteins were successfully produced via transcription and translation reactions, but the main message of the work was more general. In order to accomplish the important goal of membrane protein synthesis in SCs, the vesicle's lipid simultaneously should: (a) form good vesicles; (b) do not chemically interfere with the complicated molecular mechanism of protein synthesis; (c) allow the correct insertion and folding of the membrane protein. Therefore, a careful design is needed for functionalizing SCs with membrane proteins. There are really several relevant functions done by membrane proteins. One of the most important is the energy production. Attempts to reconstitute parts of the ATP synthase machinery have been published [23]. By coupling bacteriorhodopsin and ATP synthase [6], it has been reported the reconstitution of cytomimetic polymer vesicles, Fig. 3b, which were able to transduce light into chemical energy, i.e., ATP.

2.2 The Population/Community Perspective

This is probably the most fascinating scenario for future SC research, namely the transition from studying individual, non-interacting SCs to a population of mutually interacting ones. The population can be composed of SCs that are free moving, immobilized, networked, and so on, so that one could imagine SC communities, biofilm-like assemblies, or 2D/3D synthetic tissues (Fig. 4). Interesting work has been reported by [18], who assembled vesicles arrays by means of controlled linking (this was achieved by functionalizing the vesicle surface with nucleic acids). Protein synthesis inside such vesicle network has been also published [17]. In the context of origin of life, a simple way to observe formation of "colonies" has been reported [4]. 3D synthetic cell systems have not been announced yet, to the best of our knowledge. Spontaneous organization or directed construction could lead to 3D SCs arrays with featured properties. Ideally, a guided assembly could be realized via 3D printing, possibly on a microsculptured scaffold, and the whole system stabilized by inter-vesicles links or by a gel. Giant vesicles randomly embedded in a gel matrix are routinely prepared in our lab, displaying excellent stability. Clearly, a population of objects becomes interesting when there is a kind of physical or chemical interaction among the parts. The synthetic biology approach allows us to go one step further, foreseeing interactions that are based on exchange of *information*. This consideration brings us straightly to the next scenario.

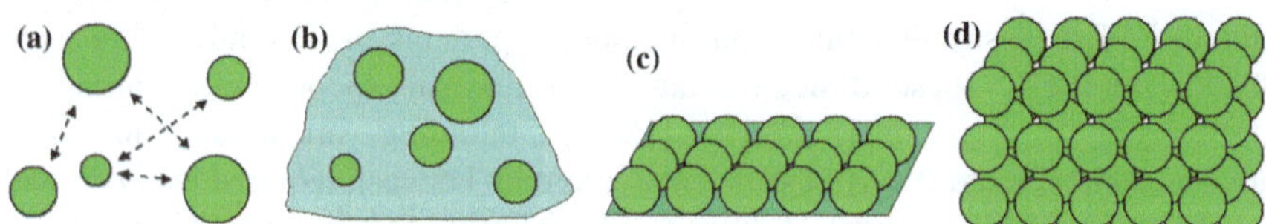

Fig. 4 From one to many. Possible experimental approaches involving synthetic cell arrays. **a** In the population/community perspective, great emphasis is given to vesicle diversity (in size, in content, in reactivity) and to the mutual interaction between them, which can be physical or chemical. **b** One easy way to immobilize synthetic cell is their embedding in a gel matrix. Alternatively, 2D (**c**) or 3D (**d**) arrays can be envisaged. These systems can be models of tissues, can be used as biotechnology tool, or can be studied to simulate primitive cells communities

2.3 Manipulation of (Bio)Chemical Signals: The Bio-Chem-ICTs Perspective

SCs constructed according to the synthetic biology paradigm (standardization of parts, modularity, orthogonality, etc.) not only can serve to first demonstrate that life can emerge from inanimate matter—e.g., consider the first man-made autopoietic cell—but also have the tremendous potential of communicate with each other. This will become possible if SCs will be able to manipulate chemical signals, performing molecular communications [33]. This innovative field can result in novel information and communication technologies based on bio-chemical signals, or bio-chem-ICTs [1]. European networks focus on these issues (cobra-project.eu, fet-circle.eu).

Inspired by previous work [8, 14, 20, 25], we have clearly defined in which way synthetic cells can be used as tools for bio-chem-ICTs [50], and research is in progress in our laboratory to create a communication protocol allowing synthetic and natural cells communicating with each other (Fig. 5a), based on quorum sensing bacterial mechanisms. In particular, our first goal is the creation of synthetic cells capable of communicating with bacteria (Rampioni et al., in preparation), but also with cells of superior organisms. SCs that communicate with biological cells could be used as programmable devices for nanomedicine, as depicted in Fig. 5b.

Another intriguing direction refers to mimicking complex morphogenetic patterns, based on the confinement of chemical oscillators in microcompartments [42]. For instance, the Belousov-Zhabotinsky reaction has been reconstituted inside

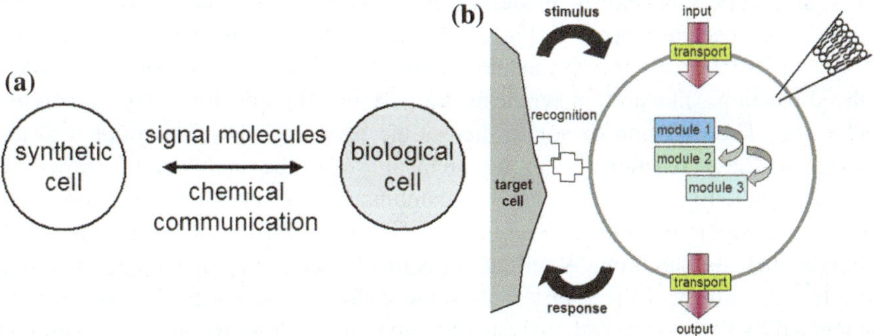

Fig. 5 a Is it possible to create a chemical communication protocol for establishing synthetic communication between synthetic and natural cells? Quorum sensing molecules and mechanisms can be exploited to establish a two-way communication channel (synthetic cells/bacteria). Taken from [50]. **b** By advancing the current biotechnology of synthetic cell construction, medical potential applications of more sophisticated, communicating and programmable synthetic cells can be envisaged. A goal would be to construct cell-like systems that, once injected in human body, reach a specic target region and thanks to chemical information processing, is able to act accordingly, for example producing in situ a cytotoxic drug or a stimulus to trigger a cellular response. Very recently, the concept of "pseudo-cell factories" or "nanofactories" (illustrated above) has been proposed. Redrawn, with minor modifications, after [25]

vesicles within liposomes [43, 51, 53]. The overall reaction is driven by the oxidation of an organic substrate, e.g. malonic acid, by bromate in acidic solution in the presence of a catalytic specie in the form of an organo-metal complex such as ferroin (a phenanthroline-iron(II) complex). The oscillatory dynamics is governed by the amount of the inhibitory intermediate bromine and the excitatory intermediate bromous acid present in the vesicles which might diffuse between individual units, thus affecting the overall oscillatory synchrony of multiple drop arrays. These intermediates actually serve as messenger molecules between individual units. In these works it has been proven that networks of oscillating vesicles provide a model to apprehend the propagation and processing of chemical information and the BZ reaction was used as "signal transmitter/receiver" to probe and study complex communication networks, mainly displaying an inhibitory character, i.e., anti-phase oscillations and stationary "Turing" structures. In fact, synchronisation of dynamic elements via chemical communication is a widespread phenomenon in nature and in relevant fields such as biology, physics and chemistry, where coupling and synchronisation are achieved by messenger molecules diffusing from one element to others triggering and spreading a chemical reaction.

2.4 The Integration of Mathematical Models

Engineering inspires synthetic biologists. Such attitude requires the integration of wet-lab approaches with mathematical modelling. Due to the very nature of SCs, which are microscopic compartments often filled with a limited number of molecules, stochastic biochemical modelling plays a major role, but the deterministic approach has been also applied (Fig. 6). We can distinguish models that operate on SCs at different levels. First, models can describe just the mechanisms of internalized reactions, like protein synthesis, transcriptional regulation, enzyme activity, and so on. The diffusion of solutes across the membrane is often coupled to the course of these reactions. The hierarchical vesosome architecture shown in Sect. 2.1 and Fig. 3a also needs a combination of chemical reactions and diffusion/transport modelling. Second, the focus can be on the vesicle transformations such as shrinking, swelling, growth, division, fragmentation, including vesicle formation and the physics of solute capture. These aspects can be complemented by the mechanical analysis of membranes (elasticity, curvature, bending energy,). Third, modelling SCs in a population context (strictly speaking, any systems with more than one SC) considers interaction among vesicles (with exchange of chemical signals), coordinated behaviour, associations in clusters, formation of high-order structures (2D/3D arrays). It is clear that in the more complex cases, a correct model can be obtained only if it spans through different scales, from the molecular nanoscale to the compartment mesoscale and above, blending together chemical and physical processes.

An important SC topic concerns stochastic effects. The SC volume lies in the aL-fL-pL range, implying that solutes are often present in a *small* number. For

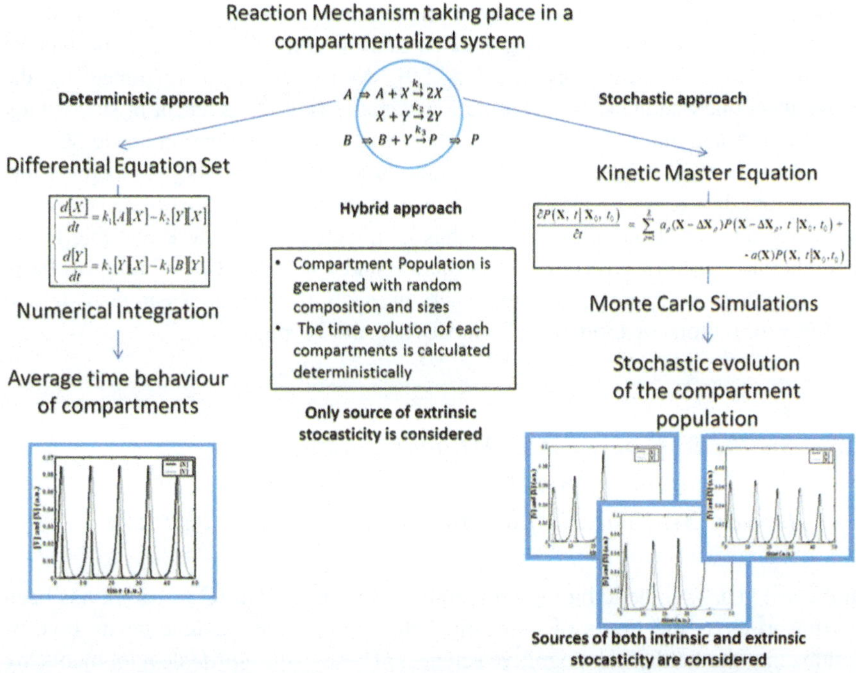

Fig. 6 Typical approaches used in theoretical modelling of compartmentalized reactions (reactions occurring inside vesicles). The specific example is on compartmentalized oscillating chemical systems (i.e., a Lokta-Volterra-like mechanism including five species A, B, X, Y, P, fuelled by external addition of the components A and B, to which the vesicle membrane is supposed to be permeable, as it is for the product P). On the *left*, the 'deterministic approach', generally used in bulk chemical kinetics, makes use of rate laws and ordinary differential equations, which are solved analytically or integrated numerically to find the time-dependent concentrations of all species, known the initial conditions. Such an approach is best suitable for systems with high number of molecules and assumes that all the reacting compartments behave similarly in average. On the *right*, the 'stochastic approach' which expresses the rate law by means of the so-called kinetic master equation, which is solved numerically via Monte Carlo simulations [15, 30]. In contrary to the deterministic approach, the stochastic approach uses the number of molecules as variables, instead of their concentration, and reaction probability instead of reaction rate. Such an approach is generally used for systems with low number of molecules. However, it should be remarked that in some specific examples (e.g., reactive systems with consumption) samples with low numerosity can be correctly modelled by a deterministic approach; and vice versa, some high numerosity systems need the stochastic approach (e.g., the repressilator [56]). More in general, the two approaches especially differ for ill-conditioned systems

instance, in vesicles of 100 nm (diameter), a concentration of 10 μM corresponds to an average of 3 molecules/vesicle. Stochastic phenomena occur both during the formation of vesicles, generating vesicles with different solute content. In turn, these vesicles will originate a kind of "diversity" in terms of the rates of internalized reactions (extrinsic stochastic effects).

We have recently studied both extrinsic and intrinsic stochastic effect in protein-synthesizing SCs, providing detailed analysis of the PURE system mechanism and its energy requirements [3, 24, 29]. Moreover, with reference to the above-mentioned anomalous encapsulation statistics (Fig. 2), a recent in silico study has demonstrated that the measured distribution of protein production inside SCs can be explained only if solute encapsulation (during SC formation) obeys to a power-law distribution, thus rejecting the null hypothesis of random encapsulation [31].

By combining observation and stochastic models it is therefore possible to get insights into the laws governing SC structure and reactivity. Contributing to basic understanding of compartmentalized reactions, such investigations can provide also useful information for modern cell physiology and biophysics.

2.5 Synthetic Biology and Artificial Intelligence: The Potential of Synthetic Cell Research for the Synthetic Exploration of Embodied Cognition

In the last years synthetic biology research, and in particular SC research, has been starting to cross the limits of its original disciplinary area, extending its investigations into the field of the cognitive sciences. These *avant-garde* developments, by shifting the focus of the inquiry from applicative to genuinely basic research purposes, are specifically related to the recent rise of the "Embodied Cognitive Science" (ECS) [7]. ESC directs the attention of research in cognitive sciences towards the fundamental role(s) the biological body plays in cognition. With this new center of interest in the scientific study of cognition, a strong synergy between ECS and biology is growing: a cooperative exploration of the biological groundings of cognitive processes to which, in the last years, SC research has been planning and inceptively starting to actively contribute.

From the methodological point of view, the role of SC research in the scientific exploration of "embodied cognition" relies on the *understanding-by-building* method [40], originally introduced by cybernetics and today increasingly adopted by scientists to study biological and cognitive processes [10]. Its key methodological concept is the following: to test and to develop scientific theories about the mechanism underlying life and cognition, by (i) incorporating related hypotheses in artificial functioning systems, and (ii) comparing the behaviours that these systems produce with those observable in the target natural systems.

Today this "synthetic" or "constructive" approach [10] methodologically supports the involvement of SCs in the exploration of cognition. This can be done by modelling and investigating natural cognitive processes via artifacts building. These artifacts are not limited to computer programs and robotic platforms, but include chemical artificial systems. In other words, the development of artificial intelligence

(AI) research not only through software and hardware, but also wetware models of cognitive processes.

From the theoretical viewpoint, this project finds its groundings in the idea that we can ascribe to minimal living systems (minimal cells) with elementary "cognitive" capabilities. According to this view, introduced by biology of cognition research in the last century and currently developed by radical approaches in ECS [9], these capabilities rely in the "autonomy" of minimal living systems, that is, the property they share with all biological organisms of producing and conserving themselves (their material identity) by actively maintaining a dynamical coupling with their environment [2]. This coupling is what allows us to think minimal living systems as minimally cognitive systems, as it relies on their capability of regulating their dynamics of self-production in response to environmental changes through processes of *perception* and *active reaction*. To be more explicit, perceiving some external variations as perturbations of the internal process of self-production; reacting to them through changes in the metabolic dynamics that compensate these alterations; stably associating to external variations internal patterns of self-regulation, that is, internally generated "operational meanings" allowing the systems to survive in the given environmental conditions.

This theoretical perspective offers to current SC research a significant role in "Embodied AI" [7]: studying minimal forms of embodied cognition—and possibly their evolution—by synthesizing and empirically exploring chemical models of minimal cells, and their interaction with their environment. The centres of interest of contemporary research are multiple, such as (i) exploring the role of regulation in the origins of biological cognition, (ii) understanding and designing minimal forms of collective intelligence, and (iii) studying specific cognitive functions—such as anticipation—in their minimal forms, among others [11].

3 Concluding Remarks

The combination of vesicle technology and cell-free systems is a very fecund vein in synthetic biology. Different approaches coexist, such as the construction of synthetic cells as primitive cell models, or as a tool for study biological processes thanks to simplification, or for designing and building artificial systems for biotechnology. We have firstly summarized the state-of-the-art of synthetic cell research, in particular of the so-called *semi-synthetic* approach [26, 47].

Starting from the recent achievements, it is possible to foresee directions for future studies on next generation synthetic cells, whose developments seem to be at the same time challenging and exciting.

Acknowledgements The authors thank Pier Luigi Luisi (Roma Tre University and ETH Zurich) for inspiring discussions. This work has been stimulated by our involvement in the European COST Action CM-1304 *"Emergence and Evolution of Complex Chemical Systems"* and TD-1308 *"Origins and evolution of life on Earth and in the Universe (ORIGINS)"*.

References

1. Amos, M., Dittrich, P., McCaskill, J., Rasmussen, S.: Biological and chemical information technologies. Procedia Comput. Sci. **7**, 56–60 (2011)
2. Bich, L., Damiano, L.: Life, autonomy and cognition: an organizational approach to the definition of the universal properties of life. Orig. Life Evol. Biosph. **42**, 389–397 (2012)
3. Calviello, L., Stano, P., Mavelli, F., Luisi, P.L., Marangoni, R.: Quasi-cellular systems: stochastic simulation analysis at nanoscale range. BMC Bioinform. **14**, S7 (2013)
4. Carrara, P., Stano, P., Luisi, P.L.: Giant vesicles Colonies: a model for primitive cell communities. ChemBioChem **13**, 1497–1502 (2012)
5. Chandrawati, R., Caruso, F.: Biomimetic liposome- and polymersome-based multicompart-mentalized assemblies. Langmuir **28**, 13798–13807 (2012)
6. Choi, H.J., Montemagno, C.D.: Artificial organelle: ATP synthesis from cellular mimetic polymersomes. Nano Lett. **5**, 2538–2542 (2005)
7. Clark, A.: An embodied cognitive science? Trends Cogn. Sci. (Regul. Ed.) **3**, 345–351 (1999)
8. Cronin, L., Krasnogor, N., Davis, B.G., Alexander, C., Robertson, N., Steinke, J.H.G., Schroeder, S.L.M., Khlobystov, A.N., Cooper, G., Gardner, P.M., Siepmann, P., Whitaker, B.J., Marsh, D.: The imitation game–a computational chemical approach to recognizing life. Nat. Biotechnol. **24**, 1203–1206 (2006)
9. Damiano, L.: Co-emergences in life and science: a double proposal for biological emergentism. Synthese **185**, 273–294 (2010)
10. Damiano, L., Hiolle, A., Canamero, L.: Grounding Synthetic Knowledge. In: Lenaerts, T., Giacobini, M., Bersini, H., Bourgine, P., Dorigo, M., Doursat, R. (eds.) Advances in artificial life, ECAL 2011, pp. 200–207. MIT Press, Cambridge MA (2011)
11. Damiano, L., Kuruma, Y., Stano, P.: What can synthetic biology offer to artificial intelligence (and vice versa)? BioSystems **148**, 1–3 (2016)
12. Elani, Y.: Construction of membrane-bound artificial cells using microfluidics: a new frontier in bottom-up synthetic biology. Biochem. Soc. Trans. **44**, 723–730 (2016)
13. Endy, D.: Foundations for engineering biology. Nature **438**, 449–453 (2005)
14. Gardner, P.M., Winzer, K., Davis, B.G.: Sugar synthesis in a protocellular model leads to a cell signalling response in bacteria. Nat Chem **1**, 377–383 (2009)
15. Gillespie, D.T.: A general method for numerically simulating the stochastic time evolution of coupled chemical reactions. J. Comput. Phys. **22**(4), 403–434 (1976)
16. Hadorn, M., Boenzli, E., Eggenberger Hotz, P., Hanczyc, M.M.: Hierarchical unilamellar vesicles of controlled compositional heterogeneity. PLoS ONE **7**, e50156 (2012)
17. Hadorn, M., Boenzli, E., Srensen, K.T., De Lucrezia, D., Hanczyc, M.M., Yomo, T.: Defined DNA-Mediated assemblies of gene-expressing giant unilamellar vesicles. Langmuir **29**, 15309–15319 (2013)
18. Hadorn, M., Eggenberger Hotz, P.: DNA-mediated self-assembly of artificial vesicles. PLoS ONE **5**, e9886 (2010)
19. Ishikawa, K., Sato, K., Shima, Y., Urabe, I., Yomo, T.: Expression of a cascading genetic network within liposomes. FEBS Lett. **576**, 387–390 (2004)
20. Kaneda, M., Nomura, S.i.M., Ichinose, S., Kondo, S., Nakahama, K.i., Akiyoshi, K., Morita, I.: Direct formation of proteo-liposomes by in vitro synthesis and cellular cytosolic delivery with connexin-expressing liposomes. Biomaterials **30**, 3971–3977 (2009)
21. Kita, H., Matsuura, T., Sunami, T., Hosoda, K., Ichihashi, N., Tsukada, K., Urabe, I., Yomo, T.: Replication of genetic information with self-encoded replicase in liposomes. ChemBioChem **9**, 2403–2410 (2008)
22. Kuruma, Y., Stano, P., Ueda, T., Luisi, P.L.: A synthetic biology approach to the construction of membrane proteins in semi-synthetic minimal cells. Biochim. Biophys. Acta **1788**, 567–574 (2009)

23. Kuruma, Y., Suzuki, T., Ono, S., Yoshida, M., Ueda, T.: Functional analysis of membranous fo-a subunit of F1Fo-ATP synthase by in vitro protein synthesis. Biochem. J. **442**, 631–638 (2012)
24. Lazzerini-Ospri, L., Stano, P., Luisi, P., Marangoni, R.: Characterization of the emergent properties of a synthetic quasi-cellular system. BMC Bioinform. **13**(4), S9 (2012)
25. Leduc, P.R., Wong, M.S., Ferreira, P.M., Groff, R.E., Haslinger, K., Koonce, M.P., Lee, W. Y., Love, J.C., McCammon, J.A., Monteiro-Riviere, N.A., Rotello, V.M., Rubloff, G.W., Westervelt, R., Yoda, M.: Towards an in vivo biologically inspired nanofactory. Nat. Nanotechnol. **2**, 3–7 (2007)
26. Luisi, P.L., Ferri, F., Stano, P.: Approaches to semi-synthetic minimal cells: a review. Naturwissenschaften **93**, 1–13 (2006)
27. Luisi, P.L., Allegretti, M., Pereira de Souza, T., Steiniger, F., Fahr, A., Stano, P.: Spontaneous protein crowding in liposomes: a new vista for the origin of cellular metabolism. ChemBioChem **11**, 1989–1992 (2010)
28. Luisi, P., Varela, F.: Self-replicating micelles—a chemical version of a minimal autopoietic system. Orig. Life Evol. Biosph. **19**, 633–643 (1989)
29. Mavelli, F., Marangoni, R., Stano, P.: A Simple Protein Synthesis Model for the PURE System Operation. Bull. Math. Biol. **77**, 1185–1212 (2015)
30. Mavelli, F., Piotto, S.: Stochastic simulations of homogeneous chemically reacting systems. J. Mol. Struct. THEOCHEM **771**(13), 55–64 (2006)
31. Mavelli, F., Stano, P.: Experiments on and numerical modeling of the capture and concentration of transcription-translation machinery inside vesicles. Artif. Life **21**, 445–463 (2015)
32. Morris, E., Chavez, M., Tan, C.: Dynamic biomaterials: toward engineering autonomous feedback. Curr. Opin. Biotechnol. **39**, 97–104 (2016)
33. Nakano, T., Moore, M., Enomoto, A., Suda, T.: Molecular communication technology as a biological ICT. In: Sawai, H. (ed.) Biological Functions for Information and Communication Technologies, pp. 49–86. Studies in Computational Intelligence, Springer, Berlin Heidelberg (2011)
34. Noireaux, V., Libchaber, A.: A vesicle bioreactor as a step toward an artificial cell assembly. Proc. Natl. Acad. Sci. U.S.A. **101**, 17669–17674 (2004)
35. Oberholzer, T., Albrizio, M., Luisi, P.L.: Polymerase chain reaction in liposomes. Chem. Biol. **2**, 677–682 (1995)
36. Oberholzer, T., Nierhaus, K.H., Luisi, P.L.: Protein expression in liposomes. Biochem. Biophys. Res. Commun. **261**, 238–241 (1999)
37. Oberholzer, T., Wick, R., Luisi, P.L., Biebricher, C.K.: Enzymatic RNA replication in self-reproducing vesicles: an approach to a minimal cell. Biochem. Biophys. Res. Commun. **207**, 250–257 (1995)
38. Paleos, C.M., Tsiourvas, D., Sideratou, Z.: Interaction of vesicles: adhesion, fusion and multicompartment systems formation. ChemBioChem **12**, 510–521 (2011)
39. Pautot, S., Frisken, B.J., Weitz, D.A.: Production of unilamellar vesicles using an inverted emulsion. Langmuir **19**, 2870–2879 (2003)
40. Pfeifer, R., Scheier, C.: Understanding Intelligence. MIT Press, Cambridge MA (2000)
41. Pohorille, A., Deamer, D.: Artificial cells: prospects for biotechnology. Trends Biotechnol. **20**(3), 123–128 (2002)
42. Rossi, F., Budroni, M.A., Marchettini, N., Cutietta, L., Rustici, M., Liveri, M.L.T.: Chaotic dynamics in an unstirred ferroin catalyzed Belousov-Zhabotinsky reaction. Chem. Phys. Lett. **480**, 322–326 (2009)
43. Rossi, F., Zenati, A., Ristori, S., Noel, J.M., Cabuil, V., Kanoufi, F., Abou-Hassan, A.: Activatory coupling among oscillating droplets produced in microfluidic based devices. Int. J. Unconv. Comput. **11**, 23–36 (2015)
44. Scott, A., Noga, M.J., de Graaf, P., Westerlaken, I., Yildirim, E., Danelon, C.: Cell-free phospholipid biosynthesis by gene-encoded enzymes reconstituted in liposomes. PLoS ONE **11**, e0163058 (2016)

45. Shimizu, Y., Inoue, A., Tomari, Y., Suzuki, T., Yokogawa, T., Nishikawa, K., Ueda, T.: Cell-free translation reconstituted with purified components. Nat. Biotechnol. **19**, 751–755 (2001)
46. de Souza, T., Steiniger, F., Stano, P., Fahr, A., Luisi, P.L.: Pereira Spontaneous crowding of ribosomes and proteins inside vesicles: a possible mechanism for the origin of cell metabolism. Chembiochem **12**, 2325–2330 (2011)
47. Stano, P., Carrara, P., Kuruma, Y., de Souza, T.P., Luisi, P.L.: Compartmentalized reactions as a case of soft-matter biotechnology: synthesis of proteins and nucleic acids inside lipid vesicles. J. Mater. Chem. **21**, 18887–18902 (2011)
48. Stano, P., D'Aguanno, E., Bolz, J., Fahr, A., Luisi, P.L.: A remarkable self-organization process as the origin of primitive functional cells. Angew. Chem. Int. Ed. Engl. **52**, 13397–13400 (2013)
49. Stano, P., Luisi, P.L.: Semi-synthetic minimal cells: origin and recent developments. Curr. Opin. Biotechnol. (2013)
50. Stano, P., Rampioni, G., Carrara, P., Damiano, L., Leoni, L., Luisi, P.L.: Semi-synthetic minimal cells as a tool for biochemical ICT. Biosystems **109**, 24–34 (2012)
51. Stockmann, T.J., Nol, J.M., Ristori, S., Combellas, C., Abou-Hassan, A., Rossi, F., Kanoufi, F.: Scanning electrochemical microscopy of belousovzhabotinsky reaction: how confined oscillations reveal short lived radicals and auto-catalytic species. Anal. Chem. **87**, 9621–9630 (2015)
52. Szostak, J.W., Bartel, D.P., Luisi, P.L.: Synthesizing life. Nature **409**, 387–390 (2001)
53. Tomasi, R., Nol, J.M., Zenati, A., Ristori, S., Rossi, F., Cabuil, V., Kanoufi, F., Abou-Hassan, A.: Chemical communication between liposomes encapsulating a chemical oscillatory reaction. Chem. Sci. **5**, 1854–1859 (2014)
54. Vlassov, A.V., Magdaleno, S., Setterquist, R., Conrad, R.: Exosomes: current knowledge of their composition, biological functions, and diagnostic and therapeutic potentials. Biochimica et Biophysica Acta (BBA)—Gen. Subj. **1820**, 940–948 (2012)
55. Walde, P., Wick, R., Fresta, M., Mangone, A., Luisi, P.: Autopoietic self-reproduction of fatty-acid vesicles. J. Am. Chem. Soc. **116**, 11649–11654 (1994)
56. Wilkinson, D.J.: Stochastic modelling for quantitative description of heterogeneous biological systems. Nat. Rev. Genet. **10**(2), 122–133 (2009)
57. Yanagisawa, M., Iwamoto, M., Kato, A., Yoshikawa, K., Oiki, S.: Oriented reconstitution of a membrane protein in a giant unilamellar vesicle: experimental verification with the potassium channel KcsA. J. Am. Chem. Soc. **133**, 11774–11779 (2011)
58. Yu, W., Sato, K., Wakabayashi, M., Nakaishi, T., Ko-Mitamura, E.P., Shima, Y., Urabe, I., Yomo, T.: Synthesis of functional protein in liposome. J. Biosci. Bioeng. **92**, 590–593 (2001)

Green Synthesis of Gold Nanoparticles from Extracts of *Cucurbita pepo* L. Leaves: Insights on the Role of Plant Ageing

Cristina Gonnelli, Cristiana Giordano, Umberto Fontani,
Maria Cristina Salvatici and Sandra Ristori

Abstract Environment-friendly and cost effective methods to obtain metal nanoparticles represent a major issue in modern material science. In particular, synthetic routes relying on green chemistry appear to be promising for large scale production. In this work, we prepared and characterized gold nanoparticles (AuNPs) from extracts of *Cucurbita pepo* L. leaves, which constitute an agricultural byproduct of large diffusion and abundant biomass. The investigation was carried out at different plant ages, from 1 to 4 months, and the production of nanoparticles (in term of size, shape and yield) was correlated with the concentration of chlorophylls and carotenoids in the extracts. The synthesis was carried out by using purely aqueous extracts at relatively low temperature (70 °C) and diluted solutions of $HAuCl_4$ (from 5×10^{-5}M to 10^{-3}M in the final samples) to provide for the gold precursor. TEM microscopy evidenced that lower Au(III) concentration promotes the formation of anisotropic particles and platelets, while higher concentrations favor a huge production of more monodisperse AuNPs with size in the range of 10–15 nm. In addition, the age of plants was showed to play a role in controlling the shape and size of the AuNPs. Our results open new perspectives for the control

C. Gonnelli
Department of Biology, University of Florence, Via P.A. Micheli 1, 50121 Florence, Italy

C. Giordano · M.C. Salvatici
Centre for Electron Microscopy "Laura Bonzi" ICCOM CNR, Via Madonna del Piano 10, 50019 Sesto Fiorentino, Florence, Italy

C. Giordano
Trees and Timber Institute, National Research Council of Italy (IVALSA-CNR), Via Madonna del Piano 10, 50019 Sesto Fiorentino, Florence, Italy

U. Fontani · S. Ristori (✉)
Department of Chemistry, University of Florence, Via della Lastruccia 3, 50019, Sesto Fiorentino Florence, Italy
e-mail: sandra.ristori@unifi.it

© Springer International Publishing AG 2018
S. Piotto et al. (eds.), *Advances in Bionanomaterials*,
Lecture Notes in Bioengineering, DOI 10.1007/978-3-319-62027-5_14

of shape and size of nanoparticles obtained by green methods in view of their applications in technological fields, which may range from nanocatalysis to biomedicine.

Keywords Green chemistry · Gold nanoparticles · Plant extract · Plant life cycle

1 Introduction

The field of green chemistry came into the world of research two decades ago, because of the increasing request for more sustainable processes in the chemical manufacturing. Such new technology aims at decreasing the use of toxic substances and the generation of noxious wastes, while preserving efficacy [1]. Over the same years, along with such general growing demand, another issue has raised from the market and industry claiming for a massive production of metal nanoparticles, due to their increasing application in the most disparate fields, such as catalysis, sensing, electronics, photonics and medicine. Actually, in the past years such spectacular development of nanotechnology was carried out without concern for a potential effect of the nanoparticle synthesis byproducts on human health and the environment. Nowadays, the two above-mentioned concerns have met and it is widely recognized that environment-friendly and biocompatible methods to obtain metal nanoparticles represent a major challenge in material science [2]. Consequently, green chemistry is increasingly directing its effort to meet sustainability requirements in nanotechnologies, involving the utilization of harmless capping agents, less dangerous reducing agents, and selection of environmentally acceptable solvents [3]. Therefore, substances from biological sources have become the preferred tool for the green synthesis of metal nanoparticles. Among them, plant extracts represent a promising route for the production of nanoparticles, being rich in molecules potentially able to act concurrently as reducing and capping agents [4]. In this context, procedures to obtain nanoparticles using plant extracts are easily scalable and cost-effective, when compared with the more expensive methods based on microbial processes and whole plants [5]. Moreover, using a conventional synthetic route, i.e. reactants mixed in solution and stirred at constant temperature for a definite time, allows better control on the ensuing nanoparticles.

Among the different plant materials, the use of agricultural leftovers for the synthesis of nanoparticles can be extremely attractive, because it not only gives new values to the discarded biomass but also contributes to eliminate a waste. For example, the red grape pomace from winery has been suggested for the synthesis of Au, Ag, Pd, and Pt nanoparticles [6] and the remnant water of soaked Bengal gram for the synthesis of gold nanoparticles (AuNPs) [7]. We already showed the effectiveness of water extracts from *Cucurbita pepo* L. [8], a species of large agricultural diffusion, whose abundant and nonedible biomass still has to find a suitable route of recycling. To find out the most effective procedures for obtaining control over yield and morphology of nanoparticles, here we carried out the synthesis of AuNPs in a set of ecofriendly and mild chemical conditions taking into account the two following parameters: (i) the plant age over its life cycle and (ii) different concentrations of gold precursor. Indeed, we supposed that extracts from plants at different stages of their life cycle could contain different concentrations and combinations of plant-derived reducing and capping agents, thus influencing the process of AuNP formation. In addition to such natural variability of the extract composition, we tested the effect of different concentrations of the precursor salt on the nanoparticle production: to the best of our knowledge, this is the first time that the effect of the combination of these two parameters has been investigated for the green synthesis of AuNPs. Our results show the dependence of the yield and morphology of the obtained nanoparticles on the selected parameters and open new direction toward a reliable control over the shape and size of green-synthesized nanoparticle in view of their applications in many technological fields, from nanocatalysis to biomedicine.

2 Materials and Methods

2.1 Plant Material and Preparation of Plant Extracts

Seeds of *Cucurbita pepo* L. (var. Faenza) were sown in peat soil during spring 2015 and grown at the greenhouse facilities of the Department of Biology at the University of Florence (Italy). After 1, 2 and 4 months of growth, fully expanded plant leaves were harvested and the samples were named 1M, 2M and 4M respectively. Leaves were washed three times with distilled water, blotted dry, weighed, ground in 3 ml g^{-1} fresh weight of distilled water and filtered with Miracloth (Calbiochem).

2.2 Preparation of AuNPs

The synthesis mixture was composed by 0.4 ml of extract, 1.1 ml of milli-Q grade water and 1.5 ml of HAuCl$_4$ solutions for a total volume of 3 ml. The extract was mixed with water and equilibrated at 70 °C before adding the solutions of metal salt, dropwise and under stirring, at different concentrations with final Au(III) concentrations in the reaction vessel of: 6.35×10^{-5}M, 1.27×10^{-4}M, 3.17×10^{-4}M, 6.35×10^{-4}M and 1.27×10^{-3}M. These samples, differing by Au (III) concentration, were named Au-1, Au-2, Au-3, Au-4 and Au-5, respectively. For the reference samples (Au-0), 0.15 ml of HCl 0.1M diluted to 0.5 mL were used instead of HAuCl$_4$. The heating and stirring were arrested after 30 min, and all samples were stocked in the dark at room temperature. Control AuNPs samples were also prepared with the standard procedure devised by Turkevich and Frens [9, 10] for the sake of comparison.

2.3   Visible Spectroscopy

Extinction spectra were recorded with a Perkin Elmer lambda 35 spectrophotometer. Samples were put in quartz cuvettes with a 1 cm path length and diluted with milli-Q water when the optical density was too high (i.e. when the absorbance in the range 500–600 nm was ≥ 3.0). The scan rate was 60 nm/min.

2.4   Electron Microscopy

Transmission electron microscopy (TEM) analyses were carried out on a TEM CM 12 PHILIPS, equipped with an OLYMPUS Megaview G2 camera, with an accelerating voltage of 80 keV. Samples for TEM analysis were prepared by suspending in water and deposing a drop of the obtained suspension on a Carbon Cu grid, followed by solvent evaporation. The samples with high amount of organic matrix were processed with a slightly different procedure: the grid was laid down on the surface of a drop of suspension for 1 min, and then, after solvent evaporation, it was introduced into the TEM.

2.5   Determination of Pigment Concentration in the Extracts

The absorbance of the extracts was determined at 470, 665.2 and 652.4 nm, using a Perkin Elmer lambda 35 spectrophotometer. The concentrations of chlorophyll a and b and of total carotenoids were calculated using the equations from Wellburn [11].

3   Results and Discussion

The presence of nanoparticles in the reaction mixture was firstly suggested by a change of color, which took place 2–3 min after adding the precursor to the extract solutions. The samples went from pale green to light orange or red/purple, depending on the concentration of gold precursor. Differences in the extinction spectra with respect to the reference systems, i.e. diluted extracts and pure $HAuCl_4$, were also detected after 1 h from the reaction, though they were not in the form of a well-defined plasmonic peak. The presence of a characteristic surface plasmon resonance of AuNPs in the range 500–600 nm with low intensity could be inferred, but it was masked by other signals such as the strong upturn of the visible spectra at low wavelengths due to the scattering of small objects in solution (Fig. 1). At higher wavelengths (>600 nm), the extinction spectra are dominated by the chlorophyll absorptions. Noticeably, these signals were strongly reduced, with the

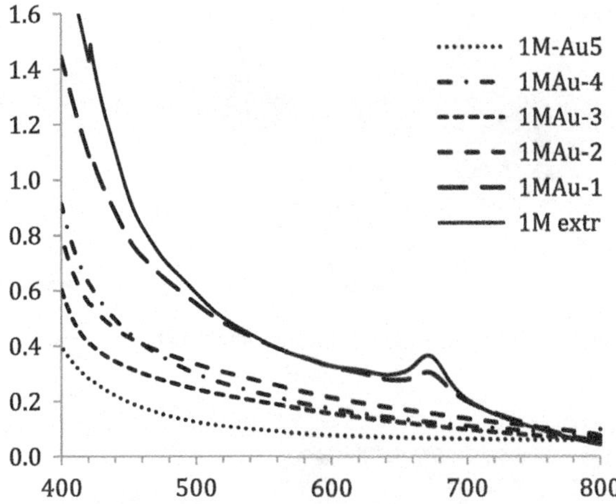

Fig. 1 Extinction spectra of the 1M reaction mixtures after 1 h from the synthesis (1M samples)

only exception of the sample containing the lowest precursor concentration, indicating that in this case the ratio Au(III)/extract was not high enough to allow all the consumption of all the chlorophylls present in solution.

Indeed, at 1 h after the synthesis, the absorbance in the spectra of the different samples was lower than that of the reference sample, but with a progressive trend which followed the increase of the Au(III) starting concentration, thus indicating the presence of nanoparticles with a high degree of polydispersity. This was subsequently confirmed by TEM analysis, as detailed in the following paragraphs. The UV-vis characterization of the reaction mixture with 2M and 4M extracts gave similar results as for samples 1M, with a decreasing absorbance of the spectra with increasing Au(III) concentration (data not shown).

3.1 Nanoparticles from Plants at the Beginning of the Life Cycle

The TEM images of 1MAu-1 samples showed few nanoparticles, most of them displaying large dimensions (100–200 nm of diameter), which were mainly of polygonal shape and often precipitated in aggregates (Fig. 2).

Increasing the Au(III) concentration, the yield of AuNPs increased, while their size progressively decreased. Moreover, the shape was predominantly spherical. Actually, in 1MAu-5 samples the size reached a minimum range of 5–15 nm (Fig. 3).

The gold nanoparticles obtained in the experimental conditions of sample 1MAu-5 were quite similar to those obtained by the method originally proposed by Turkevich [9] and refined by Frens [10], that is generally considered the reference

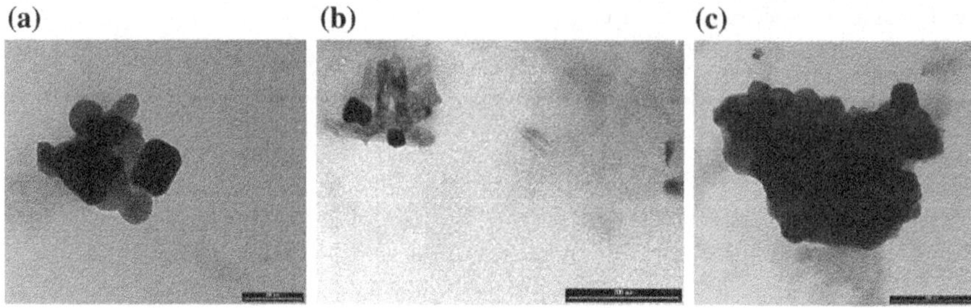

Fig. 2 TEM images obtained from sample 1MAu1. Bar scale 200 nm (**a** and **b**) and 500 nm (**c**)

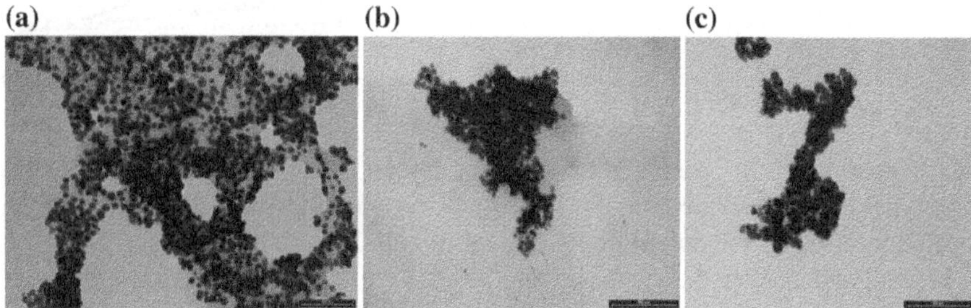

Fig. 3 TEM images obtained from sample 1MAu-5. Bar scale 100 nm

procedure to obtain AuNPs by chemical synthesis (Fig. 4). As it can be observed by comparison with Fig. 3a, the only differences with respect to particles obtained by *Cucurbita pepo* extracts are a slightly higher uniformity and the absence of an underlying organic matrix, both of which might be expected if considering the much simpler reducing agent (i.e. sodium citrate) of the Turkevich-Frens method.

In 2M samples with low Au(III) concentrations, the TEM analysis evidenced few nanoparticles, with very variable shape, even if mainly polygonal. Their size was large and reached about 800 nm for platelet-shape aggregates (Fig. 5). In this latter case, the observed particles could be possibly classified more as standard crystallites, rather than nanoparticles.

Increasing the concentration of Au(III), the number of nanoparticle increased and from sample 2MAu-3 onward the increasing presence of small spherical nanoparticles with diameter under 10 nm was observed, while the rod-like platelets were scarcely detected (see for example 2MAu-4 sample in Fig. 6).

With respect to 1M samples, in 2M systems the presence of small spherical nanoparticles was remarkable from samples Au-3 and not only from sample Au-4 onwards.

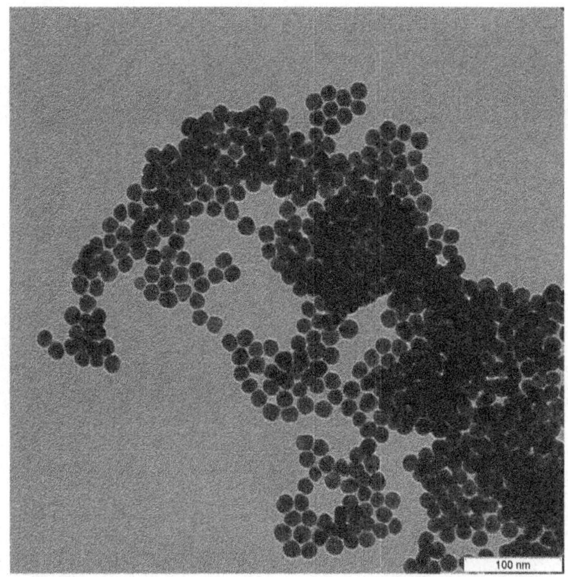

Fig. 4 TEM image of AuNPs obtained by the standard Turkevich-Frens method from $HAuCl_4$ 1.5×10^{-3}M and Na citrate 5×10^{-3}M; T = 70 °C. Bar scale 100 nm

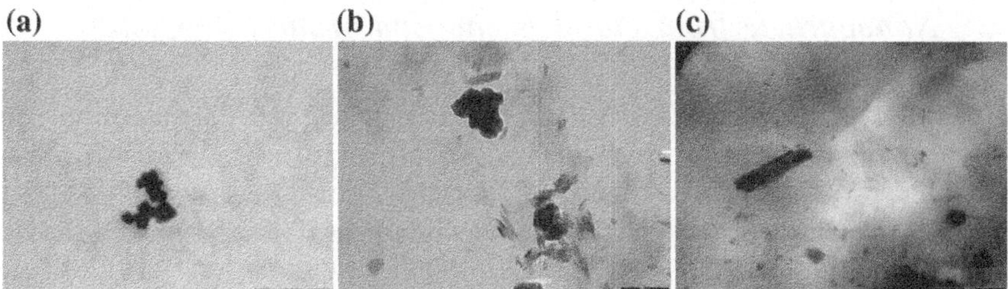

Fig. 5 TEM images of sample 2MAu-2. Bar scale 100 nm (**a**), 200 nm (**b**) and 500 nm (**c**)

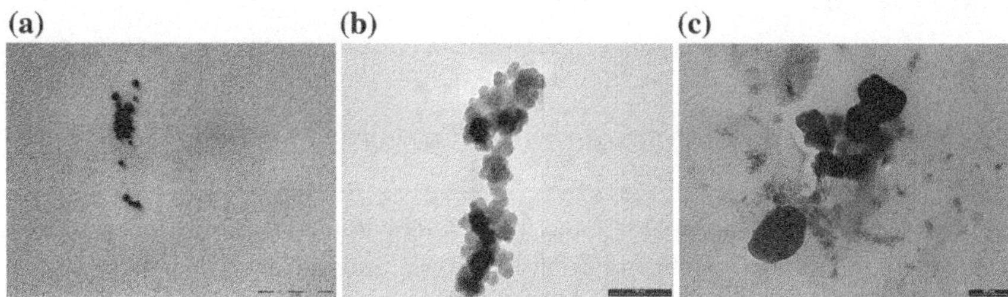

Fig. 6 TEM images of sample 2MAu-4. Bar scale 100 nm (**a**) and 200 nm (**b** and **c**)

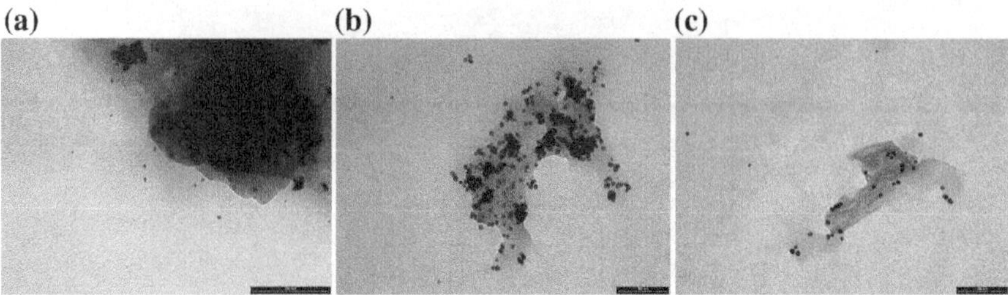

Fig. 7 TEM images of sample 4MAu-3. Bar scale 200 nm (**a**) and 100 nm (**b** and **c**)

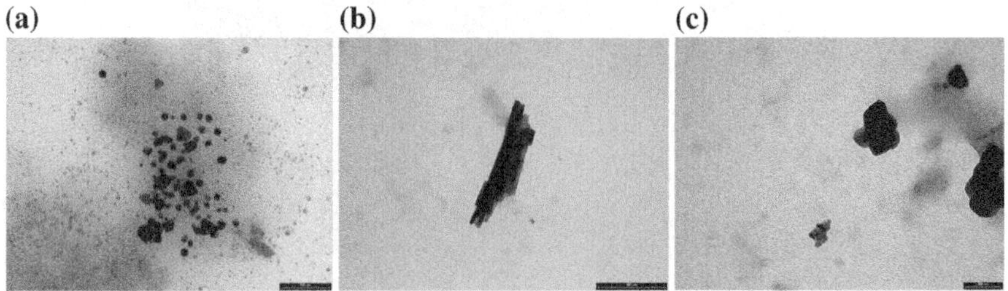

Fig. 8 TEM images of samples 4MAu-5. Bar scale 200 nm

3.2 Nanoparticle from Plants at the End of the Life Cycle

In the case of 4M samples, the TEM analysis highlighted different results in comparison with the other nanoparticles obtained using extracts from younger plants. Actually, in Au-1, Au-2 and Au-3 samples TEM images showed the presence of small nanoparticles (with a diameter lower than 10 nm) with a low degree of polymorphism and with a relatively low abundance (see Fig. 7 for 4 MAu-3 sample).

In sample 4MAu-4 and 4MAu-5, with the highest concentrations of Au(III), a copious yield of large nanoparticles with a high degree of polymorphism was showed (Fig. 8).

3.3 Pigment Concentration in the Extracts

The concentration of chlorophyll a and b and total carotenoids in the plant extracts is reported in Table 1. Both chlorophyll a and chlorophyll b increased in the extracts according to the age of the plant, whereas the concentration of total carotenoids did not follow a definite trend over the plant life cycle, thus rendering impossible any speculation on the role of these latter pigments in the synthesis of AuNPs.

Table 1 Concentration of pigments in *Cucurbita pepo* L. extracts from plant at different age (expressed as $\mu g \ ml^{-1}$ of extract). Values are means of three replicates $\pm$ standard deviation, significant differences between the means (analyzed by Tukey test) appear with different letters (at least $p < 0.05$)

	Chlorophyll a	Chlorophyll b	Total carotenoids
1M	$3.27 \pm 0.05a$	$3.42 \pm 0.09a$	$0.06 \pm 0.01a$
2M	$6.65 \pm 0.02b$	$4.68 \pm 0.01b$	$0.13 \pm 0.01c$
4M	$11.46 \pm 0.05c$	$8.15 \pm 0.02c$	$0.08 \pm 0.01b$

The peculiar result obtained with extracts from 4M plants, with production of small nanoparticles at lower Au(III) concentration, could be explained by the higher chlorophyll concentration in these extracts. In fact, due to their highly negative redox potential [-1.1–1.3 V, 12] chlorophylls can act as reducing agents for Au(III) in the presence of visible light and promote fast formation of Au(0) nuclei. These are subsequently capped by extract components, which limits their growth and results in a fairly homogeneous population of small globular AuNPs. At higher precursor concentrations, the ratio between Au(III) and plant extracts increased becoming unfavorable for efficient capping. This generated larger aggregates and occasional formation of platelets, which can be considered boundary structures between nanoparticles and crystallites.

4 Conclusions

It is well-established that plant extracts can be used as efficient reactants to obtain metal nanoparticles; in particular, *Cucurbita pepo* L. aqueous extracts have been used for such purpose [8, 13]. Here, we highlighted not only the importance of the precursor concentration, but also, and for the first time, the role played by plant growth in controlling the morphology and yield of AuNP production. Depending on the type of desired nanoparticles, optimal conditions of plant ageing and Au (III) concentration can be found by suitable adjustments of these parameters. In particular, the most abundant production of nearly spherical AuNPs, similar to those obtained by standard chemical methods, was found to occur for samples aged one month.

References

1. Collins, T.J.: Green Chemistry. Macmillan Encyclopedia of Chemistry. Simon and Schuster Macmillan, New York (1997)
2. Akthar, M.S., Panwar, J., Yun, Y.S.: Biogenic synthesis of metallic nanoparticles by plant extracts. ACS Sustain. Chem. Eng. **1**, 591–602 (2013)

3. Duan, H., Wang, D., Yadong, L.: Green chemistry for nanoparticle synthesis. Chem. Soc. Rev. **44**, 5778–57792 (2015)
4. Mittal, A.K., Chisti, Y., Banerjee, U.C.: Synthesis of metallic nanoparticles using plant extracts. Biotechnol. Adv. **31**, 346–356 (2013)
5. Iravani, S.: Green synthesis of metal nanoparticles using plants. Green Chem. **13**, 2638–2650 (2011)
6. Baruwati, B., Varma, R.S.: High value products from waste: grape pomace extract—A three-in-one package for the synthesis of metal nanoparticles. Chem. Sus. Chem. **2**, 1041–1044 (2009)
7. Ghule, K., Ghule, A.V., Liu, J.Y., Ling, Y.C.: Microscale size triangular gold prisms synthesized using bengal gram beans (Cicer arietinum L.) extract and HAuCl4·3H2O: a green biogenic approach. J. Nanosci Nanotechnol. **6**, 3746–3751 (2006)
8. Gonnelli, C., Cacioppo, F., Giordano, C., Capozzoli, L., Salvatici, M.C., Colzi, I., Del Bubba, M., Ancillotti, C., Ristori, S.: *Cucurbita pepo* L. extracts as a versatile hydrotropic source for the synthesis of gold nanoparticles with different shapes. Green Chem. Lett. Rev. **8**, 39–47 (2015)
9. Turkevich, J., Stevenson, P.C., Hillier, J.: A study of the nucleation and growth processes in the synthesis of colloidal gold. Discuss. Faraday Soc. **11**, 55–75 (1951)
10. Frens, G.: Controlled nucleation for the regulation of the particle size in monodisperse gold suspensions. Nature (London), Phys. Sci. **241**, 20–22 (1973)
11. Wellburn, A.R.: The spectral determination of chlorophylls a and b, as well as total carotenoids, using various solvents with spectrophotometers of different resolution. J. Plant Physiol. **144**, 307–313 (1994)
12. Ishikita, H., Loll, B., Biesiadka, J., Saenger, W., Knapp. E.W.: Redox potentials of chlorophylls in the photosystem ii reaction center. Biochemistry **44**, 4118–4124 (2005)
13. Chandran, K., Song, S., Yun, S.I.: Effect of size and shape controlled biogenic synthesis of gold nanoparticles and their mode of interactions against food borne bacterial pathogens. Arab. J. Chem. (2014). doi:10.1016/j.arabjc.2014.11.041

Photosynthesis Without the Organisms: The Bacterial Chromatophores

Emiliano Altamura, Fabio Mavelli, Francesco Milano
and Massimo Trotta

Abstract Anoxygenic photosynthetic bacteria are an extremely old form of life that inhabits planet Earth since approximately 3 Gya. They represent the model system for the far more complex photosynthetic organisms appeared later in time, i.e. cyanobacteria, algae and plants, and capable of performing the oxygenic photosynthesis. In this chapter we wish to present a short review of the cell architecture of one specific phototrophic bacterium, namely *Rhodobacter sphaeroides*, devoting particular attention to cytoplasmic membrane and its invagination originating under anoxygenic conditions and exposure to light. The structure and enzyme organization of the so-called chromatophores of two strains of *Rhodobacter sphaeroides* are presented, along with the isolation and purification procedures.

Keyword Photosynthetic bacteria

1 Introduction

Aliens landing on Earth will immediately realise that virtually the sole energy source for our planet is solar radiation. Next, they will realize that photosynthetic organisms are able to transduce solar radiation in other forms of energy readily viable for all life forms and, consequentially, sustain life on Earth. Their attention will be drawn toward awesome organisms like trees, bushes, or grass fields but will hardly find any interests in the small and often disregarded ecological niches where photosynthetic bacteria live and prosper.

E. Altamura · F. Mavelli
Dipartimento Di Chimica, Università Degli Studi "Aldo Moro" Di Bari,
Bari, Italy

F. Milano · M. Trotta (✉)
IPCF-CNR Istituto Per I Processi Chimico Fisici, Consiglio Nazionale Delle Ricerche,
Bari, Italy
e-mail: massimo.trotta@cnr.it

© Springer International Publishing AG 2018
S. Piotto et al. (eds.), *Advances in Bionanomaterials*,
Lecture Notes in Bioengineering, DOI 10.1007/978-3-319-62027-5_15

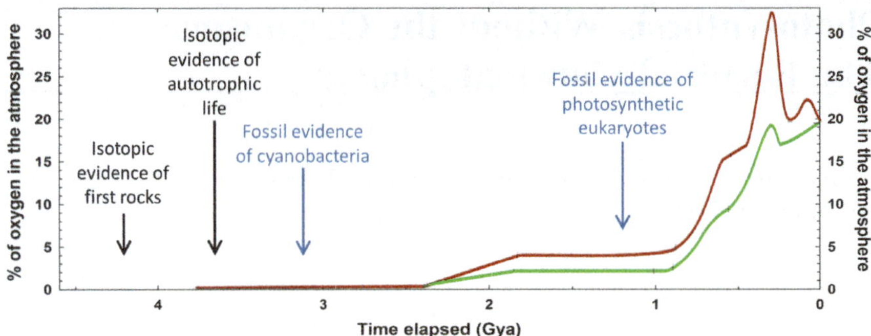

Fig. 1 The evolution of oxygen content in the atmosphere since the formation of planet Earth obtained by isotopic evidence. *Red* and *green lines* represent the higher and lower estimation of Oxygen concentration. Figure adapted from [3]

Photosynthetic bacteria "thrive in the anaerobic part of aquatic environments; from moist and muddy soils to ditches, pools, ponds, lake, sulphur springs, and marine habitats" [1, 2]. What a pity for an alien to miss them, since they will not discover some colourful spots and, most of all, they will not learn how photosynthesis works in anoxygenic conditions, i.e. the environmental conditions present on planet Earth in the first billion or so years after its formation [3, 4] (see Fig. 1).

Photosynthetic bacteria live in anaerobic environment and hence make no use of oxygen during the synthesis of the constituents of their bacterial cell. The overall chemical equation for the photosynthetic reaction taking place in anoxygenic bacteria was described by Van Niel [5] in the form:

$$CO_2 + 2H_2A + \text{Light Energy} \rightarrow [CH_2O] + 2A + H_2O \qquad (1)$$

where H_2A represents any electron donor molecule able to reduce carbon dioxide and eventually fix it. The above equation is quite general and remains valid for oxygenic photosynthetic organisms; in the latter case A represents oxygen.

2 Classification of Photosynthetic Bacteria

Photosynthetic bacteria are classified on the bases of their distinctive features in five main families, according to Fig. 2.

The first branching of the scheme deals with the electron donor species of Eq. 1: photosynthesis in cyanobacteria is based on two large enzymatic complexes called Photosystem I and II; the latter includes the so-called oxygen evolving complex, a protein subunit were electrons withdrawn from water are used in the bacteria metabolism. The overall reaction produces oxygen as by-product. The appearance of these microorganisms on Earth is responsible for the dramatic modification of the planet atmosphere shown in Fig. 1 and, as consequence, for the emergence of all oxygen-based life forms.

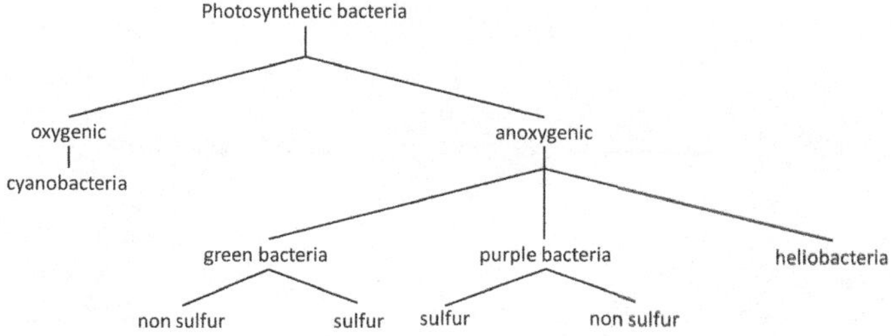

Fig. 2 Classification of the photosynthetic bacteria

Green phototrophs are gram-negative bacteria and fall in two categories; the first collects the nonmotile and obligate anaerobe collectively known as Chlorobiaceae. They depend on the presence of light and hydrogen sulphide to develop. Hydrogen sulphide is reduced to sulphate or, depending on the intensity of light and on the sulphide concentration, to sulphur that collects in small granules outside the bacterial cells. The second category is represented by the Chloroflexaceae, filamentous gliding bacteria, able to thrive under anoxygenic conditions, performing photosynthesis mostly using organic compounds as electron donor, and under oxygenic conditions. Both categories have a photosynthetic apparatus related to the photosystem I of the cyanobacteria.

Purple bacteria are gram-negative motile species and are classified according to their ability to oxidize sulphur to sulphate. In particular, all purple sulphur bacteria do show such capability, while the purple non sulphur photosynthetic bacteria cannot withdraw electrons from sulphur [6]. As result, purple non sulphur photosynthetic bacteria either directly oxidize sulphide to sulphate or, if sulphide reduction produces elementary sulphur, the latter is deposited outside the cells and left unused. Purple non sulphur photosynthetic bacteria exhibit a notable range of morphological, biochemical and metabolic diversity. Purple bacteria develop a photosynthetic apparatus related to the photosystem II of the cyanobacteria.

The first member in the group of Heliobacteria was identified in the soil of a garden in the campus of Indiana University, Bloominton in 1983 [7]. Remarkably this newly discovered *Halobacterium chlorum* belongs to the class of gram-positive bacteria, differently to all other phototrophs. Subsequently several other members of the Heliobacteria were identified and cultured in several laboratories. Heliobacteria grow photosynthetically over a limited range of organic compounds and present a photosynthetic apparatus related to the photosystem I of the cyanobacteria.

3 The Wild Type of the Bacterium *Rhodobacter sphaeroides*

The purple non sulphur bacteria *Rhodobacter* (*R.*) *sphaeroides*, main actor of the present article, is taxonomically placed in the α subgroup of Proteobacteria and exhibits a very versatile metabolic capability, being able to grow

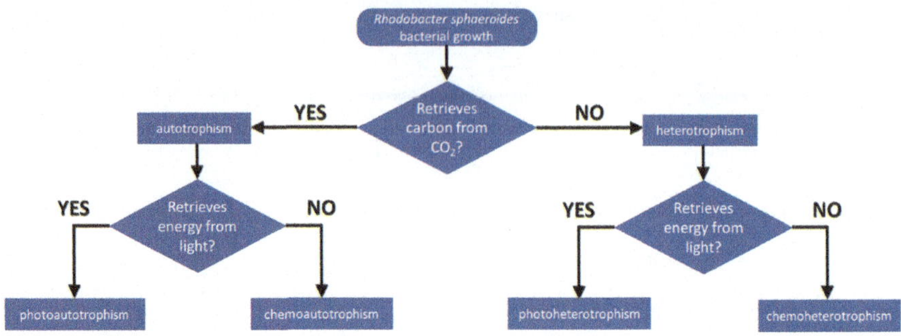

Fig. 3 A flowchart description of the metabolic versatility of the purple non sulphur bacterium *Rhodobacter sphaeroides*

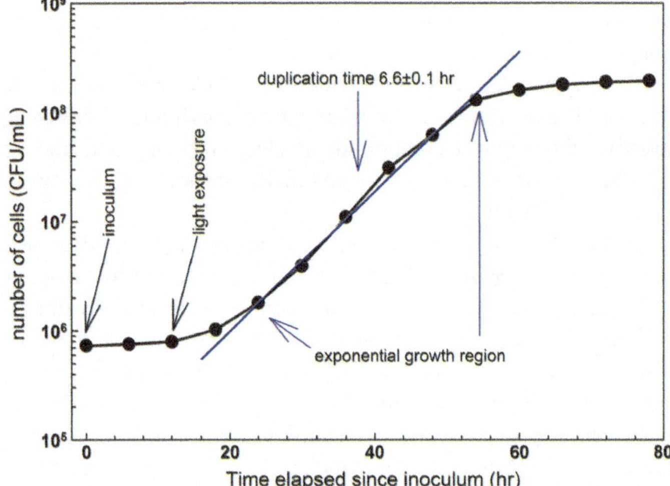

Fig. 4 Growth curve for a culture of *R. sphaeroides* wild type exposed to 10^4 lx in a closed system. Redrawn from [8]. The duplication time of the strain R26 (see text) is twice larger

photoautotrophically, photoheterotrophically, chemoautotrophically, and chemoheterotrophically according to the scheme in Fig. 3.

Rhodobacter sphaeroides is able to use a variety of electron donors and carbon sources during its photoheterotropic growth, namely acetate, citrate, ethanol, fructose, fumarate, gluconate, glucose, glycerol, lactate, malate, mannitol, pyruvate, succinate, tartrate, yeast extract.

Cell culture of the wild type strain of *R. sphaeroides* under illumination and kept in anoxygenic conditions grow photosynthetically. A typical culture requires an initial inoculum followed by a dark adaptation time that cells will use to consume all oxygen present in the liquid medium by respiration. Six to twelve hours after the inoculum, the cell culture is exposed to light (in the range between 1500 and 65,000 lx) and the photosynthetic growth initiates, entering in the exponential growth phase (see Fig. 4).

Photosynthetically grown cultures assume a deep red colour because of the presence of carotenoids molecules, are rod-shaped and their dimensions are 0.5–0.8 μm in width and 1–2 μm in length [9]. The cellular architecture shows the presence of several components:

- An external cell wall responsible for the maintaining the cellular shape and to ensure osmotic protection to the inner cell. The thickness of the cell wall ranges between 5–6 nm [10];
- An internal membrane that separates the periplasm from the cytoplasm. The thickness of inner membrane ranges between 7–8 nm [10].
- The intracytoplasmic membranes (ICM), invaginations of the membrane that forms exclusively under illumination and in anaerobic conditions and that contains the entire photosynthetic apparatus. These ICM can be obtained mechanically isolated from the cell membrane. Under opportune conditions, isolated ICM form closed spherical vesicles and can be purified to high degree. These small vesicles were firstly isolated and identified in 1952 and named chromatophores [11, 12] as they appear strongly coloured due to the presence of the photosynthetic pigments bacteriochlorophylls and carotenoids.
- A single flagellum medially located externally to the cell that allows to swim and change direction [13, 14]. A video showing swimming cells of *R. sphaeroides* is available at https://www.youtube.com/watch?v=wcJIK7EP7UE.

4   The Energy Conversion Apparatus

R. sphaeroides is capable to grow chemoheterotrophically in presence of oxygen. Reducing the partial pressure of O_2 to less than 3% (compared to 21% of the atmosphere) in the culture vessel and illuminating, imposes the transition toward photosynthetic metabolism, triggering the invagination of the cytoplasm membrane, to form the ICM, and the consequent synthesis and assembly of the photosynthetic apparatus. The yield in bacterial biomass is rather high, reaching 4 g of wet cells per litre of liquid medium. Once the mature photosynthetic growth conditions are reached, the cascade of reactions starting with the absorption of light and eventually ending with the synthesis of the energy-rich adenosine triphosphate (ATP) will be responsible for powering the entire cellular metabolism. The photon-to-ATP conversion apparatus is mainly based on a cluster of four main proteins sketched in Fig. 5:

- The light harvesting complexes (LHC), a set of two transmembrane multi-subunit proteins organized in circular shape, responsible of the light harvesting and its subsequent transfer to the following enzymatic complex;
- The photosynthetic reaction center (RC), a three-subunit transmembrane enzymatic complex, that functions as the photochemical core where photons are used to reduce ubiquinone-10 (coenzyme Q) into ubiquinol withdrawing electron from the redox protein cytochrome c_2 (cyt c_2) present in the periplasm. As consequence of these reactions, protons are sequestered from the cytoplasm and stored into the membrane;

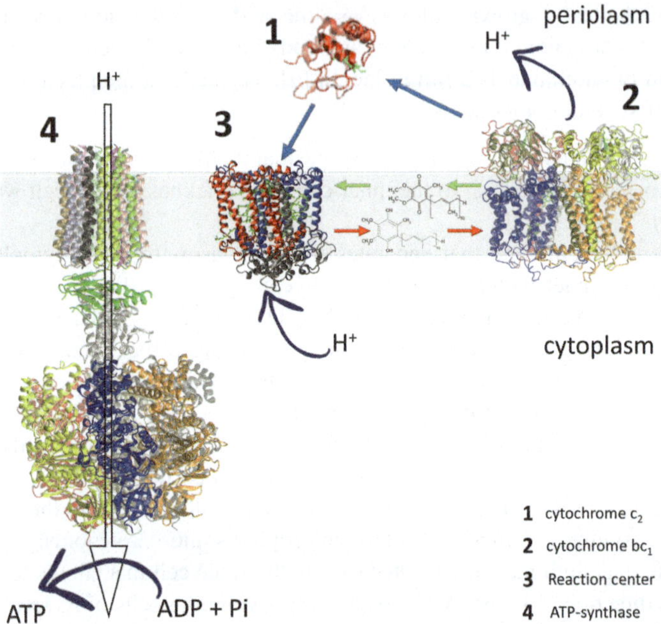

Fig. 5 Schematic representation of the main components of the photosynthetic and proton translocating apparats of a purple non sulphur bacterium. Sitting in the center of the IC membrane are shown the molecule of ubiquinone-10 and ubiquinol-10. Enzymes and molecules are drawn not in scale. Red arrows indicate the release of ubiquinol by the RC and its uptake by the cytochrome bc_1. *Green arrows* indicate the release of ubiquinone by the cytochrome and the subsequent uptake by the RC. *Blue arrows* indicate the transfer of one electron from the cytochrome bc_1 to the oxidized cyt c_2, and from the reduced cyt c_2 to the RC. For sake of clarity, the light harvesting complexes are omitted. See also [15, 16]

- The ubiquinol-cytochrome c oxidoreductase (cyt bc_1), a multi-subunit membrane enzyme, which uses the ubiquinol molecule produced by the RC to oxidize the cyt c_2. During this reactions protons are released into the periplasm;
- The net ΔpH established between periplasm and cytoplasm will be eventually used to drive the ATP-synthase, a multi-subunit enzyme having a large membrane and a soluble portions, which translocates protons back from the periplasm to the cytoplasm and synthesizes ATP from ADP and inorganic phosphate.

5   The Chromatophores

The contemporary exposure to anaerobic conditions and to light triggers the invagination of the cytoplasmic membrane and the development of the photosynthetic apparatus, which is exclusively located into the ICM. The isolation of the

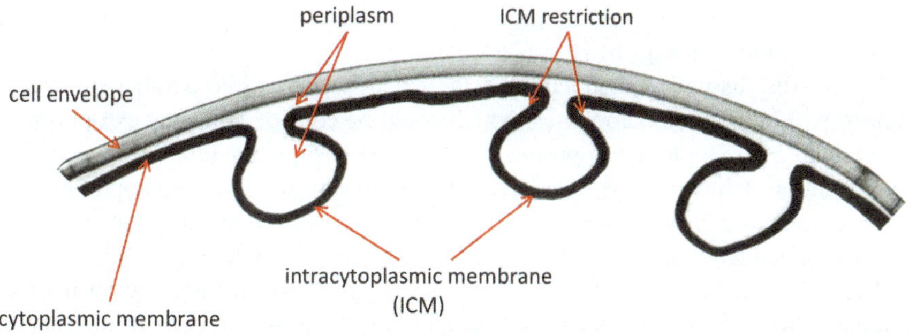

Fig. 6 Representation of a portion of the cell from *R. sphaeroides*

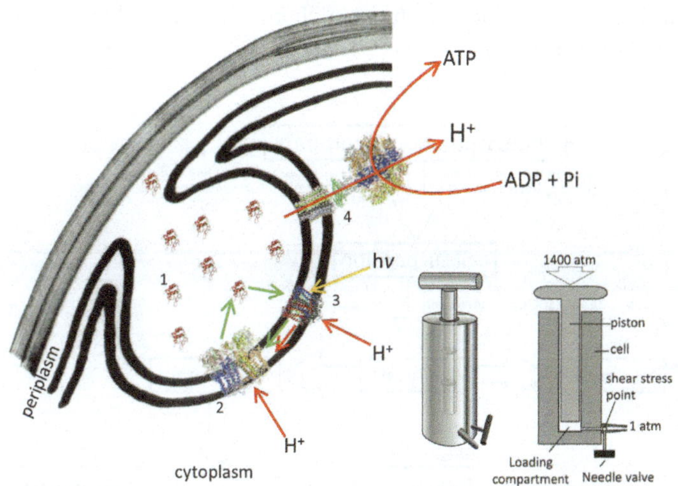

Fig. 7 Zoom of a single ICM. The orientation of the enzymatic complexes illustrated in Fig. 5 is shown. Cyt c_2 is confined in the periplasm, while the synthesis of ATP takes place in the cytoplasm. Numeration of enzymes is coincident to Fig. 5. In the inset is shown the French Pressure Cell used to disrupt the cell wall of photosynthetic bacteria retaining the activity of enzymes and producing chromatophores. This press consists of a cell with a central bore on the *top* and a *smaller* bore at the bottom. The bore at the *bottom* is sealed with a needle valve. The cell suspension is poured in the upper bore with the lower one closed. A piston is than inserted and a pressure of 1400 atm is applied. When the needle valve is opened, the suspension is squeezed through and experience a shear stress and a sudden decompressions. Bacterial cells are efficiently disrupted and the biological material retains full activity if all operations are performed at temperature lower than 4 °C

ICM requires a method to disrupt the bacterial cell wall that is accomplished mechanically by the use of a French Pressure Cell Press (see inset Fig. 7).

Luckily enough, the use of the cell press ensures the full activity of the enzymes present in the bacteria and, in the specific case of *R. sphaeroides*, results in the closure of the ICM that detach from the membrane and reorganize in closed vesicle, the chromatophores. The sealing process most probably takes place at the neck of the invagination (see Fig. 6) [11, 12].

Upon cells disruption, a fraction of the cyt c_2 is lost while another portion is trapped within the chromatophores, which could be considered as the cytoplasm of the vesicle (Fig. 7). On the opposite, the ATP-synthase will synthesize ATP in the open solution, which can be considered the periplasm of the chromatophores. For these reasons, the energy transduction and proton translocation in chromatophores are often considered as inverted compared to the intact whole cell [17].

The chromatophores have a diameter varying from 60–100 nm and an internal volume of roughly 1.2 μL per mg of protein [17]. The overall content in reaction centers depends on the actual size of the vesicle; for vesicles having a diameter of 60 nm, Saphon et al. found an average of 20 reaction center and 2–3 ATP-synthase per chromatophore [18]. Both enzymes are fully active [19, 20].

The purification procedure of chromatophores present in literature requires 7 steps and is shown in Fig. 8.

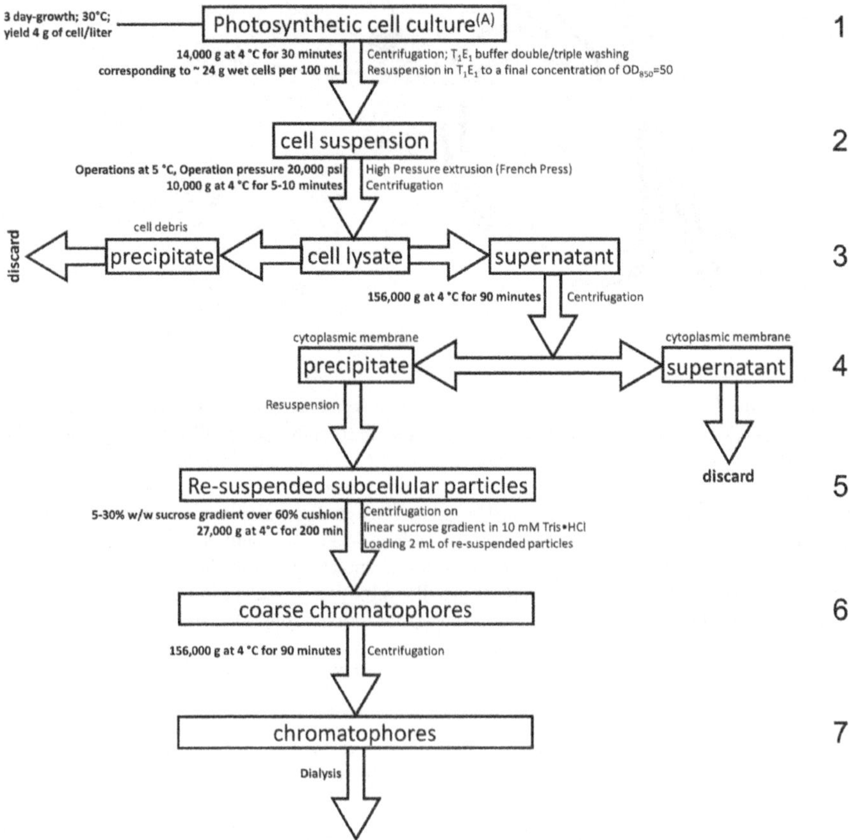

Fig. 8 The procedure for preparation and purification of chromatophores

Fig. 9 Stained thin-section of a cell form R. sphaeroides R26 examined under microscope with a ×99,000 magnification. Two kinds of intracyoplasmic membranes are visible, the spherical (**S**) ones and the lamellar (**L**) ones. The measuring bar corresponds to 0.1 μm. The inset shows a 27,000 magnification factor. Reproduced from [21] with permission

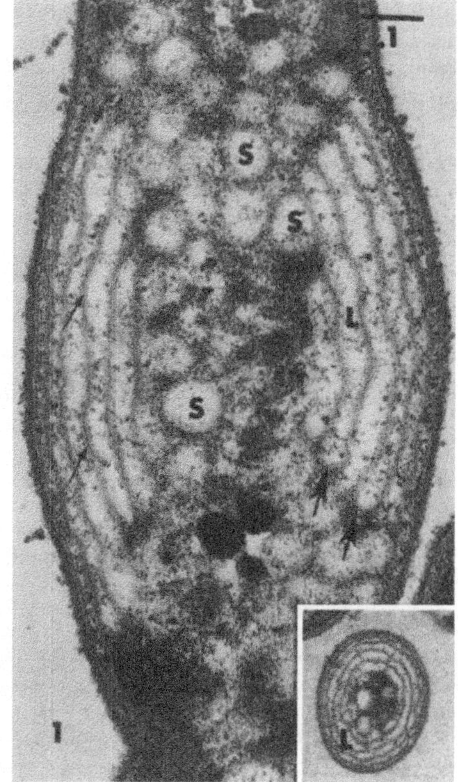

6 The Strain R26 of *R. Sphaeroides*

A green coloured carotenoidless mutant of *R. sphaeroides* named R26 is also deeply investigated as model system in bacterial photosynthesis. Not much is available in literature on the ICM organization in this strain. Lommen and Takemoto [21] showed that, differently from the wild type described above, the ICM of *R. sphaeroides* R26 are not homogenously developed, but form a inhomogeneous assembly of round and lamellar investigations, as shown in Fig. 9.

7 Conclusions

Purifyed chromatophores represent a very intriguing system to be used in almost in-vivo investigations, including the photochemistry of the reaction center and the synthesis of ATP by ATP-synthase. Chromatophores offer to chase the possibility of isolate the sole and entire photosynthetic apparatus, getting rid of all the "obnoxious" biological components not involved in the process of light transduction. The short review here presented cannot be considered in any way exhaustive. State-of-the-art investigations on the structure of ICM and on the architecture of the enzymatic organization within the chromatophores are present in literature [10, 22].

Acknowledgements MT wishes to thank Prof. Benoît Leblanc, Université de Sherbrooke (Canada) for the scheme of the French press cell reproduced in Fig. 7.

References

1. Koblizek, M.: Ecology of aerobic anoxygenic phototrophs in aquatic environments. FEMS Microbiol. Rev. **39**(6), 854–870 (2015). doi:10.1093/femsre/fuv032
2. Pfennig, N.: Photosynthetic Bacteria. Annu. Rev Microbiol. **21**(1), 285–324 (1967). doi:10.1146/annurev.mi.21.100167.001441
3. Hohmann-Marriott, M.F., Blankenship, R.E.: Evolution of Photosynthesis. Annu. Rev. Plant. Biol. **62**(1), 515–548 (2011). doi:10.1146/annurev-arplant-042110-103811
4. Milano, F., Trotta, M.: Give me a photon to stand on, and I will synthesize life! In: Macagnano, A., RamundoOrlando, A., Farrelly, F.A., Petri, A., Girasole, M. (eds.) Advanced Topics in Cell Model Systems, pp. 49–52. Nova Science Publishers Inc, Hauppauge (2009)
5. van Niel, C.B.: On the morphology and physiology of the purple and green sulphur bacteria. Arch. Mikrobiol. **3**(1), 1–112 (1932). doi:10.1007/bf00454965
6. Madigan, M.T., Jung, D.O.: An overview of purple bacteria: systematics, physiology, and habitats. In: Hunter, C.N., Daldal, F., Thurnauer, M.C., Beatty, J.T. (eds.) The Purple Phototrophic Bacteria, pp. 1–15. Springer, Dordrecht (2009). doi:10.1007/978-1-4020-8815-5_1
7. Gest, H., Favinger, J.L.: Heliobacterium chlorum, an anoxygenic brownish-green photosynthetic bacterium containing a "new" form of bacteriochlorophyll. Arch. Microbiol. **136**(1), 11–16 (1983). doi:10.1007/bf00415602
8. Zeng, X.H., Roh, J.H., Callister, S.J., Tavano, C.L., Donohue, T.J., Lipton, M.S., Kaplan, S.: Proteomic characterization of the Rhodobacter sphaeroides 2.4.1 photosynthetic membrane: identification of new proteins. J Bacteriol **189**(20), 7464–7474 (2007). doi:10.1128/Jb.00946-07
9. Cohen-Bazire, G.: Some observations on the organization of the photosynthetic apparatus in purple and green bacteria. In: Gest, H., San Pietro, A., Vernon, L.P. (eds.) Bacterial Photosynthesis. Charles F. Kettering Research Laboratory. Antioch Press, Yellow Springs, Ohio (1963). doi:10.5962/bhl.title.7230
10. Scheuring, S., Nevo, R., Liu, L.N., Mangenot, S., Charuvi, D., Boudier, T., Prima, V., Hubert, P., Sturgis, J.N.: Reich Z (2014) The architecture of *Rhodobacter sphaeroides* chromatophores. Biochim. Biophys. Acta **8**, 1263–1270 (1837). doi:10.1016/j.bbabio.2014.03.011
11. Pardee, A.B., Schachman, H.K., Stanier, R.Y.: Chromatophores of *Rhodospirillum rubrum*. Nature **169**(4294), 282–283 (1952)
12. Schachman, H.K., Pardee, A.B., Stanier, R.Y.: Studies on the macro-molecular organization of microbial cells. Arch. Biochem. Biophys. **38**, 245–260 (1952)
13. Armitage, J.P.: Swimming and behavior in purple non-sulfur bacteria. In: Hunter, C.N., Daldal, F., Thurnauer, M.C., Beatty, J.T. (eds.) The Purple Phototrophic Bacteria, pp. 643–654. Springer, Dordrecht, (2009). doi:10.1007/978-1-4020-8815-5_32
14. Armitage, J.P., Macnab, R.M.: Unidirectional, intermittent rotation of the flagellum of *Rhodobacter sphaeroides*. J. Bacteriol. **169**(2), 514–518 (1987)
15. Tangorra, R.R., Antonucci, A., Milano, F., la Gatta, S., Farinola, G.M., Agostiano, A., Ragni, R., Trotta, M.: Hybrid interfaces for electron and energy transfer based on photosynthetic proteins. In: Handbook of Photosynthesis, 3rd edn. Books in Soils, Plants, and the Environment. CRC Press, pp 201–220 (2016). doi:10.1201/b19498-14

16. Ruggiero, M., Savoia, M.V., Giotta, L., Agostiano, A., Milano, F., Trotta, M.: Solubilization of functionally connected cytochrome bc(1) complex and reaction center from *Rb sphaeroides* R-26.1 chromatophores by beta-dodecylmaltoside. BBA-Bioenergetics **1658**, 170 (2004)
17. Michels, P.A.M., Konings, W.N.: Structural and functional properties of chromatophores and membrane vesicles from *Rhodopseudomonas sphaeroides*. Biochim. Biophys. Acta **507**(3), 353–368 (1978). doi:http://dx.doi.org/10.1016/0005-2736(78)90346-2
18. Saphon, S., Jackson, J.B., Lerbs, V., Witt, H.T.: The functional unit of electrical events and phosphorylation in chromatophores from *Rhodopseudomonas sphaeroides*. Biochim. Biophys. Acta **408**(1), 58–66 (1975)
19. Elema, R.P., Michels, P.A., Konings, W.N.: Response of 9-aminoacridine fluorescence to transmembrane pH-gradients in chromatophores from *Rhodopseudomonas sphaeroides*. Eur. J. Biochem. **92**(2), 381–387 (1978)
20. Lundin, A., Thore, A., Baltscheffsky, M.: Sensitive measurement of flash induced photophosphorylation in bacterial chromatophores by firefly luciferase. FEBS Lett. **79**(1), 73–76 (1977)
21. Lommen, M.A.J., Takemoto, J.: Ultrastructure of carotenoid mutant strain R-26 of *Rhodopseudomonas sphaeroides*. Arch. Microbiol. **118**(3), 305–308 (1978). doi:10.1007/bf00429122
22. Bahatyrova, S., Frese, R.N., Siebert, C.A., Olsen, J.D., van der Werf, K.O., van Grondelle, R., Niederman, R.A., Bullough, P.A., Otto, C., Hunter, C.N.: The native architecture of a photosynthetic membrane. Nature **430**(7003), 1058–1062 (2004). doi:http://www.nature.com/nature/journal/v430/n7003/suppinfo/nature02823_S1.html

Index

© Springer International Publishing AG 2018
S. Piotto et al. (eds.), *Advances in Bionanomaterials*,
Lecture Notes in Bioengineering, DOI 10.1007/978-3-319-62027-5

The manufacturer's authorised representative in the EU is Springer
Nature Customer Service Centre GmbH, Europaplatz 3, 69115 Heidelberg,
Germany. If you have any concerns regarding our products, please
contact ProductSafety@springernature.com

Printed and bound by CPI Group (UK) Ltd, Croydon, CR0 4YY
28/11/2025
02007692-0015